Architectures of Weaving

From Fibers and Yarns to Scaffolds and Skins

Christiane Sauer / Mareike Stoll /
Ebba Fransén Waldhör / Maxie Schneider

jovis

Table of Contents

Designing Performance

Fiber Structures

Fabricating Space

Entangled Environments

Appendix

Preface

Christiane Sauer, Mareike Stoll

In September 1957 Anni Albers (1899–1994), textile artist and influential weaver, teaching at Bauhaus Dessau and Black Mountain College, wrote with regards to Architecture and Weaving:

> If the nature of architecture is the grounded, the fixed, the permanent, then textiles are its very antithesis. If, however, we think of the process of building and the process of weaving and compare the work involved, we will find similarities despite the vast difference in scale. [...] The essentially structural principles that relate the work of building and weaving could form the basis of a new understanding between the architect and the inventive weaver. New uses of fabrics and new fabrics could result from a collaboration; and textiles, so often no more than an afterthought in planning, might take a place again as a contributing thought.[1]

Today, seventy-five years later, her outlook on bridging the disciplines of architecture and textile gains new momentum as we stand at a threshold in search of a new fabric—a fabric of thought that meets the challenges of our time like climate change and resource scarcity. We are further in search of a tangible fabric for our built environment that has to become more sustainable, more intelligent, and more circular. Could we challenge the idea of *firmitas*[2] as a statically materialized architectural setting and instead replace it with a textile approach that is resilient and robust precisely because it is flexible and reconfigurable? During the last pandemic years, concepts of architecture that seemed set in stone were questioned and have become fluid as spaces have become hybrid and stretched out into the virtual. Rooms have adapted to scenarios of concurrence, interactivity, and multitude. Such an attempt to achieve more with less brings to the fore the need for a new and adaptive materiality.

The building sector today is the main global consumer of energy, both in the construction of new architecture and the operation of existing structures.[3] The demand will further increase with a growing global population. Novel material strategies play a key role when aiming at a more sustainable building industry. In the field of architecture, an intense search for novel processes and materials has already begun. In materials science, on the other hand, "programmable" or "architected" materials are being developed with customized properties based on their engineered inner geometry. They transfer the idea of bionics—a technical device based on principles found in nature—into the very structure of a material or a surface. Thus, materials themselves become active and can behave like a folding leaf or an expanding sponge. Textiles will play a central role when developing such "intelligent" material systems since fiber

---

1 A. Albers, "The Pliable Plane: Textiles in Architecture," in *Anni Albers. Selected Writings on Design*, ed. B. Danilowitz (Middletown: Wesleyan University Press 2000), 44–51, here 44 and 51.

2 Vitruvius (Roman architect and theorist, ca. 80–15 BCE) identified the three columns of architecture as *firmitas, utilitas, venustas* (strength, utility, and beauty) in his treatise *De architectura* (30–15 BCE), which is regarded as the first book on architectural theory.

3 In 2020, the energy consumption of buildings and construction accounted for 36% of global energy demand. The development of sustainable materials and processes is one of the key messages of the current Status Report for Building and Construction of the UN Environmental Program. To achieve the pledges of the Paris Agreement, the global buildings and construction sector must almost completely decarbonize by 2050. United Nations Environment Programme, *2021 Global Status Report for Buildings and Construction: Towards a Zero-emission, Efficient and Resilient Buildings and Construction Sector*, Nairobi.

is the basic building element in both natural-grown and man-made fabric. Fiber-based construction marks a shift from the top-down perspective of deploying a supposedly passive material for a specific purpose to the bottom-up process of creating and designing from and with the inner activity of a material.

Therefore, *Architectures of Weaving* explores weaving and textiles as a structural approach. Weaving becomes a means of constructing, while the weave is a spatial structure in itself—a three-dimensional layering of fibers and threads passing over and under and around each other. Such fiber architectures can even build up into complex fibrous scaffolds, as with multilayered weaves or spacer fabrics (Fig. 1, 2). On the other hand, building elements can be assembled into weave-like structures tracing a space open to appropriation (Fig. 3). Even ancient textile techniques like basket weaving can be scaled to inspire architectural concepts that merge material, structure, and form into a novel architectural substance (Fig. 4, 5).

Fig. 1
Multi-layered weave: three interwoven layers of Nylon and Cotton threads result in a spatial fabric structure. Henriette Ackermann, Weißensee School of Art and Design, Berlin, Master Studio C. Sauer, 2015.

Fig. 2
3D warp-knitted spacer fabric: stable yet elastic and permeable textile surface.

Fig. 3
Serpentine Pavilion, Sou Fujimoto, Kensington Gardens London 2013.
Latticework composed of fine steel bars. Fujimoto describes the delicate semi-transparent structure as a "multi-purpose social space" and "a transparent terrain that encourages people to interact with and explore the site in diverse ways."

Architects keep dreaming about changeable spaces that adapt to the environment and to varying needs. If clothing as our second skin is designed to be adaptive (taking on and off layers, opening and closing jackets) why is the third skin—the envelope of a room or building—still so inactive and static? Facades today try to shield us from a presumably "hostile" environment with insulation, sealings, and air conditioning. But what if the elements—sun, wind, rain—cease to be perceived as enemies in the quest for comfort, instead become sparring partners for harvesting energy or water from the skin of our buildings? What if the activity of growth were incorporated into structural concepts? What if even other species were invited for co-habitation?

Fig. 4
Basket Weaver using kagome triaxial weaving technique, Japan, ca. 1915.

Fig. 5
Haesley Nine Bridges Golf Club House, KACI + Shigeru Ban Architects, Yeoju, South Korea, 2010.
Kagome shaped wooden roof construction covering an area of 36 × 72 meters.

Textile construction is an open system that allows for such concepts. But it requires new tools, skills, and techniques rather than classical building equipment. Fiber structures can be fabricated by one person handcrafting, or by complex textile machinery operated by highly skilled personnel. Even robotic or drone-driven construction is now being tested in this context. Experimentation is inherent since textile construction is never an off-the-shelf process. Its complex performance is hard to calculate in advance, although considerable steps are being made with the latest computation technology — both on the textile and on the architectural scale.

Pioneers of textile construction in architecture like Antoni Gaudí, Frei Otto, or Heinz Isler worked long before the possibilities of computational design were even on the horizon. They generated their structures through hands-on experimentation and material-based "form-finding" with the help of physics and the flexibility of membranes or threads. Gravity shaped the catenary geometries of loaded strings in the upside-down hanging models of Antoni Gaudí's designs (Fig. 6). Surface tension in Frei Otto's soap-film experiments inspired the development of double curved canopies for large-scale roofing (Fig. 7). Elastic rubber membranes and frozen fabrics gave Heinz Isler fundamental insights when designing his ultra-thin and large-span concrete shells.[4]

Fig. 6
Upside-down hanging model for the design of Crypta of *Colonia Güell*, Barcelona, 1898-1914 by Antoni Gaudí. The catenary chains are shaped by gravity and weight to explore the ideal structural shapes.

4   Antoni Gaudí first deployed the structural tool of hanging models for the Crypta of *Colonia Güell* (1898-1914) and transferred the principle to the *Basilica de la Sagrada Família* (1882-ongoing), Barcelona. Frei Otto realized his visionary tensile constructions with the West German Pavilion at the Expo 67, Montreal, and the roof of the Olympic Stadium, Munich, 1972, amongst others. Heinz Isler proposed three basic methods of textile form-finding in his paper "New Shapes for Shells": "the freely shaped hill, the membrane under pressure and the hanging cloth reversed." H. Isler, "New Shapes for Shells," *Bulletin of the International Association for Shell Structures*, no. 8 (1961): 123–30.

Fig. 7
West German Pavilion at the Expo '67 in Montreal under construction, Rolf Gutbrod and Frei Otto, 1967. The tensioned steel cable-net covers an area of 8,000 square meters by using a double curved geometry that is fixed at only 8 steel posts and surrounding ground anchors.

Beyond explorations into structural form, the softness and drapability of the material itself has inspired architects to create novel concepts of use and perception. Lilly Reich freed the curtain from being a window covering and made it an autonomous architectural element—as a soft linear and curved gesture, an undulating wall encircling space. In her collaboration with Ludwig Mies van der Rohe for the installation of *Café Samt und Seide* (Berlin, 1927) and *Villa Tugendhat* (Brno, 1928–30) the hanging and even moveable partitions became room-scaled swatches of precious material like silk or velvet, introducing the active physical presence and sensu-

Fig. 8
Women's Fashion Exhibition *Die Mode der Dame*, Berlin, Germany. View of the *Café Samt und Seide*, designed by Ludwig Mies van der Rohe and Lilly Reich, 1927. Gelatin silver print. Mies van der Rohe Archive, gift of the architect. Acc. n.: AD531.

Fig. 9
Curtain Wall House,
Shigeru Ban Architects,
Tokyo, Japan, 1995.
Two-story high curtains can
be opened and closed by the
user to change the spatial and
climatic environment.

PREFACE

ousness of fabric into the static architectural realm (Fig. 8). This idea of textile as a means for spatial transformation was carried into contemporary architecture with Shigeru Ban's *Curtain Wall House* (Tokyo, 1995). Able to actively transform it, the inhabitants become creators of their own space. The translucent moveable curtain is rooted in Japanese building tradition, where paper-screen walls do not separate, but rather connect the inside and the outside: they are a filtering device for light, sound, and air to animate the interior environment with the elements of nature (Fig. 9).

A materialized extension of the body that traces the actions of its inhabitants instead of forcing them into a prescribed geometry is Frederick Kiesler's concept of the *Endless House* that he developed in the 1920s. He created the terms of *Correalism* and *Biotechnik*[5] for his novel approach to design spaces in flux—with walls, floors, and ceilings merging as looped elements. His sketches entangle lines like threads that condense into a shell-type enclosure while being drawn (Fig. 10). This resembles construction strategies that can even be found in other species—like silk worms when spinning cocoons (Fig. 11, 12). While Kiesler relied on model building with cladded wire mesh for showcasing his ideas, the rise of computation from the 1990s on offered the possibility to realize buildings with a freed language of form generated by CAD/CAM. Breaking open the conventions of modernism, novel architectures, fluid and textile in shape, rose with the works of architects like Frank Gehry or Zaha Hadid. Complex in form and elaborated in construction, the spatial gesture in fact remained physically static.

5 O. Oberhuber, *Frederick Kiesler, Architekt 1890–1965*, exhibition catalogue (Hochschule für Angewandte Kunst Wien, 1987), 12–13.

Fig. 10
Study for *Endless House*,
Frederick Kiesler, New York,
ca. 1959, pencil on paper.

Fig. 11–12
Colorized micro–computed
tomography (micro-CT)
images of an individual cocoon
of Epanaphe Moth spec. in
lateral view and as detail view.
Visible inside the cocoon is the
organism in its pupal stage.
The specimen was collected
in Ivory Coast by a Maninka
trader and purchased in Mali
for the "West African wild
silks" project of the Cluster of
Excellence *Matters of Activity*.

Relating to these legacies, could we today learn from the practices of silkworms and actually start building with textile as a material? What if architecture were woven? Could textiles gain more spatiality and architecture more textility? How can the idea of softness and interconnectivity change the way we perceive and use space? Can we replace the paradigm of maintaining a building fixed in its initial state with a process of re-making, re-modeling — of perpetual change? In his *The Four Elements of Architecture*, Gottfried Semper describes the textile techniques of braiding and knotting as ancient and fundamental techniques to create spatial enclosures (textile walls and screens)(Fig. 13).[6] Constructing with an entangled thread would allow for coherence and for reconfiguration — even in

6   G. Semper, *Die vier Elemente der Baukunst: ein Beitrag zur vergleichenden Baukunde* (Braunschweig: Verlag Friedrich Vieweg und Sohn, 1851), Eng. trans. *The Four Elements of Architecture and Other Writings* (Cambridge: Cambridge University Press, 1989).

Fig. 13
Braided knot, Gottfried
Semper, ca. 1860 from
G. Semper, *Der Stil in den
technischen und tektonischen
Künsten oder Praktische
Aesthetik*. Erster Band.
Textile Kunst. (Frankfurt am
Main: Verlag für Kunst und
Wissenschaft, 1860), 186.

the third dimension.[7] It would imply circularity, since the thread is kept intact after the building fabric is disassembled after use.

A weave is the materialization of its making with the hand of its maker and the traces of its tools inscribed. The rhythm of making is embedded in the structure of alternating threads and voids. The pliability of the thread allows for endless configurations in order to program a symbolic or physical "activity" into the fabric. This makes textile a means for coding through the array of stitches in a knit or crossings in a weave.[8] The common grid-notation system of binding patterns holds the information for lifting and lowering the warp yarns by filled (up) and non-filled (down) squares (Fig. 14). This binary logic of weaving became a precursor for computation. As early as the eighteenth century the drawn pattern was transferred into a grid of holes punched into paper cards for automated weaving machines. The

Fig. 14
Draft notation for plain
weave, Anni Albers, ca. 1965
from A. Albers, *On Weaving*,
(Middletown, Conn.: Wesleyan
University Press), 196.

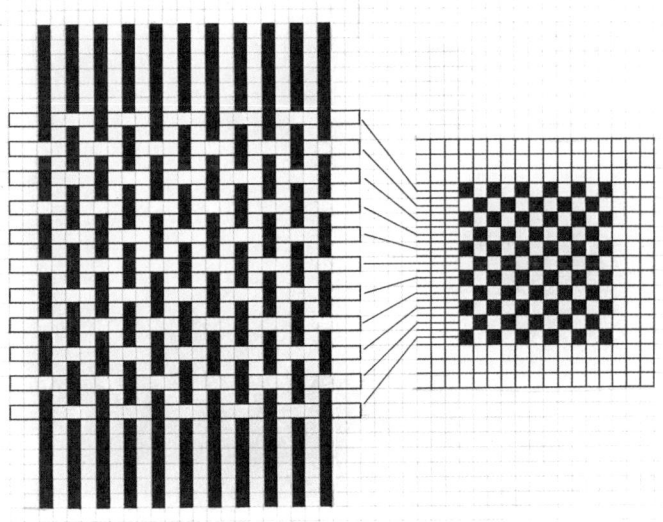

7   Architect and theorist Lars Spuybroek makes a case for a "soft constructivism" in his essay "The Structures of Vagueness." He argues to "short-circuit" Semper's distinction between the tectonic and the textile, between the spatial construction and the flat plane: "Having weak textile threads teaming up into rigid collective configurations is an upgrade or inversion of the Semperian paradigm. But they should be three-dimensional from the start, plan-threads can twist and become wall-threads. All these techniques already exist in textile art where complex interlacings occur in crochet, weaving and knitting." L. Spuybroek, "The Structure of Vagueness," *Textile: The Journal of Cloth and Culture* (January 2005): 358.
8   During World War I, knitting was even established as a means of secret coding. British intelligence gained information on German troop movements through local women in Belgium and France who encrypted information on train movements into fabric patterns through the array of the stitches—knit and purl. While crafting with needles and yarns, they acted as spies, passing on information through an innocent-looking sweater.

holes in the cards operated the shedding of the warp threads and thus created the pattern of the weave (Fig. 15). Later on, this system of punch cards was adapted for processing binary data in the very first computers that converted the physical information into digital code.[9] Even textile itself can be used as a device for storing information, as in the Khipu artifacts of the Inca Empire (ca. 1400–1533 CE). Khipus are still the subject of intense study today, because their meanings have not yet been deciphered. They consist of a system of knotted strings—a textile binary code of presumably numerical, linguistic, and even narrative content. Not only the pattern but also the physical color and materiality seem to have played a central role in communicating the embedded information (Fig. 16). With its strong bond of material, pattern, and code, it becomes obvious that textile technique is an ideal tool to program a surface in the physical and symbolic realm.

Fig. 15
Jacquard loom controlled by a continuous band of punched cards feeding into the weaving machine from the top. The punched code operates the shedding mechanism of the warp threads. The information of complex and large-scale patterns could be stored and produced in a replicable way. Engraving, published in Paris, 1876.

9   As early as 1725, Basile Bouchon, a French textile worker, invented a process to control a loom with the help of perforated paper bands. In 1804, weaver and merchant Joseph Marie Jacquard developed the first fully programmable loom with a continuous band of punch cards. Charles Babbage, an English mathematician and engineer, ideated the "Analytical Engine"—a data-processing machine based on punched paper in the 1840s. He cooperated with the mathematician and writer Ada Lovelace, who developed the algorithm and therefore became the first computer programmer. In 1890, Herman Hollerith applied the system of perforated cards for data processing in the US census. He had his invention patented and became founder of a company that was later renamed IBM (International Business Machines)—the pioneering computer-technology company.

Fig. 16
Khipu, knotted cords encoding
information. Inca Empire,
Pacasmayo Valley,
La Libertad (Peru), ca. 1430–
1530 (assumed).

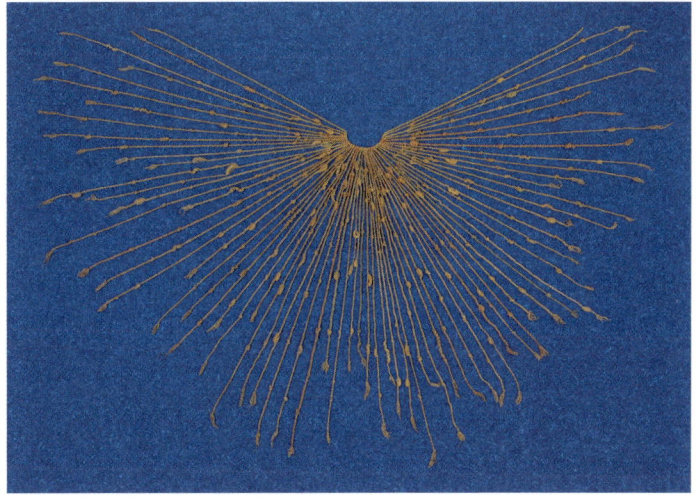

With *Architectures of Weaving*, we want to explore this inter-connectivity of material and meaning. Therefore, we look back into traditions of making and into practices that have been forgotten; we reach into micro- and nano-levels or learn from processes in time like growth and adaptation. Here, we use the term "weaving" as a synonym for all kinds of "fiber practices." This encompasses not just the crossing of warp and weft on a loom, but all textile construction techniques, like for instance, braiding or knitting, as well as theoretical approaches such as woven mathematical topologies, and political contexts such as social fabrics or even trans-species practices like bacterial weaving. Therefore, the book becomes a heterogeneous collection of contributions rooted not only in architecture and textile, but also in engineering, materials science, microbiology, mathematics, cultural science, anthropology, art and design. They are stitched together into five chapters that highlight aspects of Making, Activity, Structure, Softness, and Entanglement. We weave together perspectives from a wide variety of fields of research, rooted in both theory and practice.

Architectures of Weaving originates in the research context of the Cluster of Excellence *Matters of Activity. Image Space Material* at Humboldt-Universität zu Berlin.[10] In this context, more than forty disciplines from the fields of *Gestaltung*, the humanities and sciences work together to create a basis for a new culture of materials. Many of the contributions are rooted in the distinct research settings of *Matters of Activity*: outcomes of collaborations between cultural history, microbiology and architecture, textile design and biomaterials science, mathematics and weaving. In the context of the Cluster, an ongoing bond has established between the Weißensee School of Art and Design, Department of Textile and Surface Design, and the Max Planck Institute of Colloids and Interfaces, Department of Biomaterials, that resulted in novel formats of teaching and research on bio-inspired fiber-based design. Many of these outcomes that spark our ongoing research are showcased in the book.

Picking up the thread and taking Anni Albers' idea of the "contributing thought" literally, we—the editorial team with an interdisciplinary background in textile, architecture, and the humanities—aim to bridge disciplines to inspire new and sustainable approaches in all the fields involved. In July 2021, we organized an international

10   The Cluster of Excellence *Matters of Activity. Image Space Material* is funded by Deutsche Forschungsgemeinschaft (German Research Foundation) within the framework of the Cluster of Excellence (EXC) funding line within Germany's Excellence Strategy EXC 2025 – 390648296 (funding period 1 January 2019 – 31 December 2025).

symposium on *Architectures of Weaving* to bring together manifold perspectives with participants from within and outside the Cluster. On this day, the discussions circled around topics of scale (and scaling), the situatedness of design and knowledge, textile thinking in architecture, and textiles as a catalyst for material culture. This intense exchange is now part of the content in this book, which in itself is much like a fabric composed of many different colors and textures. The speakers' contributions have become essays and are complemented by Case Studies that were chosen as state-of-the-art approaches to the topic of "Architectures of Weaving." They encompass built examples, research prototypes, design studies, and art installations, which derive mostly from trans-disciplinary backgrounds and collaborations. We invite you—the reader—to weave your own narrative from this thread of texts and images in order to create together a fabric of novel thoughts and future perspectives.

Practices of Making

As an ancient cultural practice, the making of textiles has always been closely linked to the scale and movements of the human body. With spinning or weaving, the hand is an essential tool. It creates fabrics as protective layers encompassing not only the body itself, but also the inhabited space with woven walls, screens, canopies, or carpets. These traditions can still be found today in vernacular architecture. In most recent research, they are translated into contemporary production technologies including robots, drones or even augmented gloves.

However, the practices of making as we posit them in the following case studies are not limited to textile craft. In nature, weaving is found on many scales: in microbiology for instance, we see bacterial cells "spinning" filaments and "weaving" a tissue-like matrix for their protection. Even on an architectural scale, biological elements can be woven: in a rural area of North-East India, the roots of Rubber Fig trees are guided and entangled to form "growing" bridges. In contrast to built structures, these resilient bridges become stronger, not weaker, with time, as their joints thicken due to the living constituents. Forming a strong tensile structure from a presumed weak material is shown in an example that interleaves paper stacks into a reconfigurable bridge structure.

Three-dimensional weaving is deeply embedded in geometrical and mathematical logic. The inherent topological relations can connect seemingly distant fields: the physical practice of weaving and the theoretical practice of experimental mathematics come together and create fertile ground, as illustrated in a vivid conversation between "weavers" from both disciplines at the end of this chapter.

The Event of a Fiber

Regine Hengge, Karin Krauthausen

ONTO THE LENGTH OF A THREAD:
THE TEXTILE

Let us begin, like Anni Albers, with an event. In 1965 the artist and theorist wrote in her book *On Weaving* that her thoughts were guided by "the event of a thread."[1] This statement is notable, particularly with regard to the material set out in this article. We do not regard a thread as an everyday object or a passive material—no, a thread is an active occurrence that emerges from the everyday to become an "event." A thread attracts attention. It surprises those who engage with it. For Albers, the activity of the thread is neither a semantic abstraction nor a practical concretization. A thread does not merge inconspicuously with an existing framework; rather, it enriches it by adding something unfamiliar and, in this sense, new. The momentum emanating from a thread is tied to the thread itself; it resides within the thread, within its material structure, which becomes a driving force, an element imbued with potential (Fig. 1). It is not surprising, therefore, that Albers strongly emphasizes structure in her analysis of the woven textile:

> The structure of a fabric or its weave—that is, the fastening of its elements of threads to each other—is as much a determining factor in its function as is the choice of the raw material. In fact, the interrelation of the two, the subtle play between them in supporting, impeding, or modifying each other's characteristics, is the essence of weaving.[2]

For Albers, weaving is not the result of an active subject (the weaver) shaping a passive material into an equally passive finished product; instead, by giving rise to and limiting possibilities (to which the weaver responds creatively), it is the material, and the structures that define it, that initiate the decisions. The active/passive opposition is transformed into a configuration of "shared or distributed agency."[3]

FIBERS, FIBRILS, AND FILAMENTS:
THE FABRIC OF LIFE IN THE NANOWORLD

In nature, too, thread-like elements are spun into fibers, fibrils, and filaments on all scales, which in turn are woven into three-dimensional structures. All life on our planet is based on thread-like macromolecules, which are the fundamental components of all cells, whether single bacterial cells or human cells. In principle, DNA, RNA, proteins, and polysaccharides (the fabric that allows life to emerge and exist) are one-dimensional polymer chains that are "spun" by enzymes, that is, by energy-driven molecular machines made up of three-dimensionally folded proteins.

1   A. Albers, *On Weaving* (London: Studio Vista, 1974, first published 1965), 15.
2   Ibid., 38.
3   See É. Balibar and S. Laugier, "Art. Agency," in *Dictionary of Untranslatables: A Philosophical Lexicon*, ed. B. Cassin and, for the English translation, E. Apter et al. (Princeton and Oxford: Princeton University Press, 2014), 17-24.

THE EVENT OF A FIBER

Fig. 1
Woven Cosmos by
Hella Jongerius, exhibition
at Gropius Bau Berlin
(April 29–August 15, 2021).

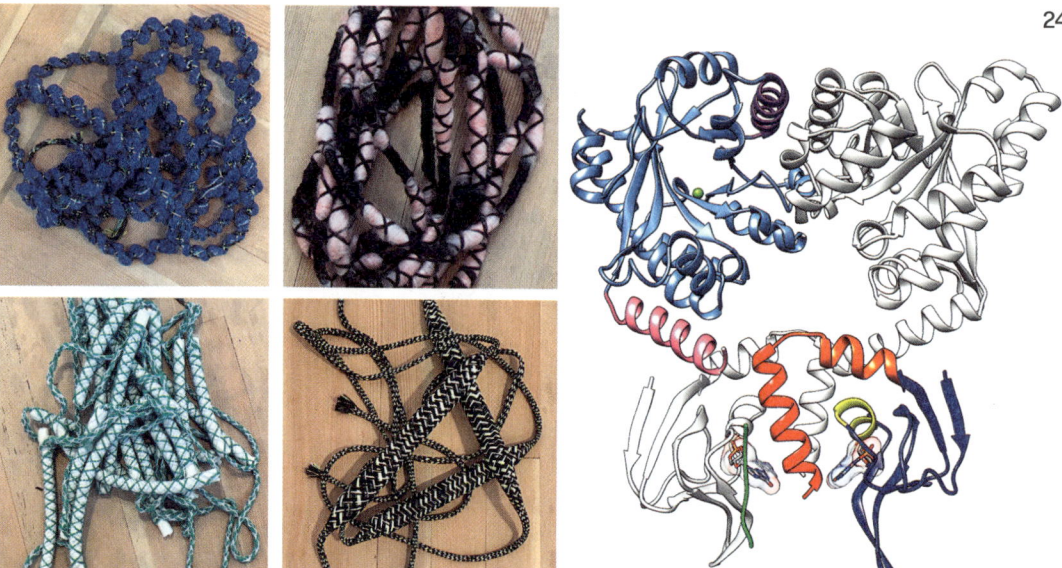

Fig. 2
Entangled fibers, fibrils and filaments. Left panel: *Woven Cosmos* by Hella Jongerius. Right panel: Structural model of a folded protein that acts as an enzyme to degrade the signaling molecule c-di-GMP.

At this smallest scale, in the nanoworld, the self-activity and self-organization of matter is the decisive factor in the assembly of functional units. The basis for this is the inherent thermal vibration that allows molecules to explore space and potential interactions—the so-called Brownian motion of molecules—and to reach deeper energy levels by means of structurally determined molecular interactions. This applies to the tangling of two DNA threads in the famous double helix, as well as to the folding of newly synthesized proteins into their final three-dimensional structures (Fig. 2), which act as cellular building blocks or molecular machines.

CELLS EMBEDDED IN EXTRACELLULAR MATRICES
OF INTERWOVEN FIBERS:
THE FABRIC OF LIFE IN THE MICROWORLD

In the living microworld, cells are the shaping actors that "spin" and "weave" molecular fibers. Bacterial cells can produce proteins and polysaccharides internally and transport them through the cell wall to their surfaces, where polysaccharide threads are spun into intertwined fibrils, which in turn are woven into two-dimensional (i.e., planar) or three-dimensional meshes that hold the cell clusters together (Fig. 3). A prominent example is the polysaccharide cellulose, which is produced by bacterial and plant cells. This is one of the most common biomolecules and a major component of wood, cotton fabrics, and paper. But also protein molecules, which have already been folded three-dimensionally, further assemble at the cell surfaces to form long fibers or filaments. These are capable of additional interactions and enable the cells to adhere to each other, that is, to mechanically glue together to form tissues. This applies in particular to the highly stable beta-amyloid protein fiber networks found, for example in silk, but also in the brains of Alzheimer's patients, where these fibers become entangled and form toxic "plaques." In bacterial biofilms, which are multicellular aggregates of bacteria that appear in many places in the environment and in our bodies, beta-amyloid protein fibers are often interwoven with cellulose fibrils. This fibrous composite material forms a stable and elastic extracellular matrix that connects the bacterial cells into a tissue-like consortium and protects them from toxic environmental influences and predators.

 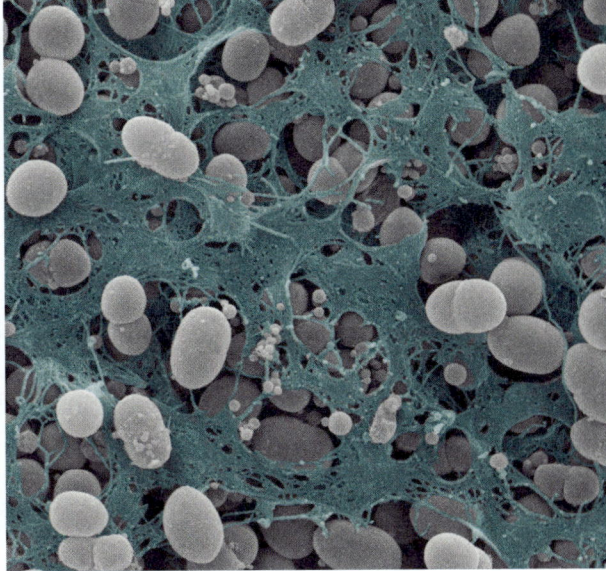

Fig. 3
Multicellular life is based on extracellular matrices of interwoven fibers. Left panel: *Woven Cosmos* by Hella Jongerius. Right panel: Scanning electron microscopic image of a bacterial biofilm. Cells of the bacterium *Escherichia coli* are surrounded by a matrix of self-produced, interwoven cellulose fibers, which provides protection and cohesion for their community. Individual cells have a diameter of about 0.001 mm. The cellulose mesh is reproduced in a false blue color.

WHY DOES LIFE BUILD WITH FIBERS INSTEAD OF BRICKS?

Using a mechanically superb fiber material consisting of entangled cellulose fibrils and amyloid protein fibers, millions of bacteria can jointly produce a striking architecture that is more than 100 times larger than the individual cells that cooperate to "weave" it (Fig. 4). From the perspective of molecular biology, highly active molecular fibers, fibrils, and threads are, therefore, what binds the living world's innermost core together. Thus, instead of using three-dimensional bricks as building blocks, nature "spins," "felts," and "weaves" by working extensively with quasi-one-dimensional fibers to assemble three-dimensional functional units on higher scales.

In relation to their volume, fibers are delicate lightweights with vast surfaces that can simultaneously form numerous weak interactions. The sum of these interactions results in strong overall cohesion, while the possibility remains that the connections can be loosened locally or reshaped without the entire tissue tearing or collapsing. This intricate, intertwined world of fibers forms soft, elastic, dynamically changeable and yet stable structures—precisely the kind of structures that active, growing, living systems need inside their cells, as well as at their flexibly expandable interfaces and protective envelopes, where controlled communication with their environment takes place.

Fig. 4
The three-dimensional weaving of the "extended organism." Left panel: Fluorescence microscopic image of a vertical thin section through a bacterial biofilm of *Escherichia coli*. The fluorescent green color is due to a dye (Thioflavin-S) that stains the extracellular matrix, which consists of tightly interwoven cellulose fibrils and amyloid protein fibers. Individual bacterial cells can be seen as tiny black dots in this stained matrix architecture. The entire matrix layer has a height of about 0.05 mm. Right panel: Three-dimensional weaving in *Woven Cosmos* by Hella Jongerius.

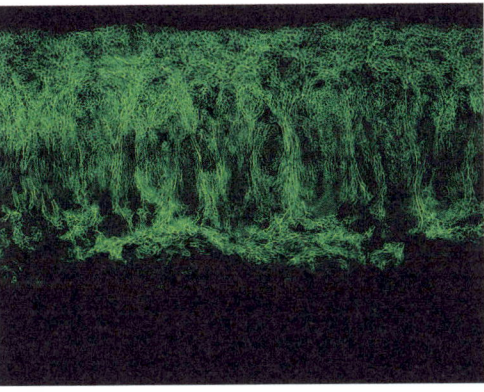 

In bacterial biofilms the interwoven fiber architecture around the cells (Fig. 4) transforms the extracellular space, which would be unpredictable and dangerous for an individual cell, into a protected intra-biofilm space that the bacteria can control homeostatically and which thereby becomes life-friendly. Essentially functioning as a "city of microbes," the biofilm establishes a common supply structure for the bacteria as an extension of the latter's own individual (cell) bodies. This kind of extended organism can also be found among social insects such as termites, whose impressive constructions far exceed the size of any individual animal. This behavior is described by the physiologist J. Scott Turner in his book *The Extended Organism: The Physiology of Animal-Built Structures*, which shows how collective structures can be beneficial to an individual organism's life.[4]

 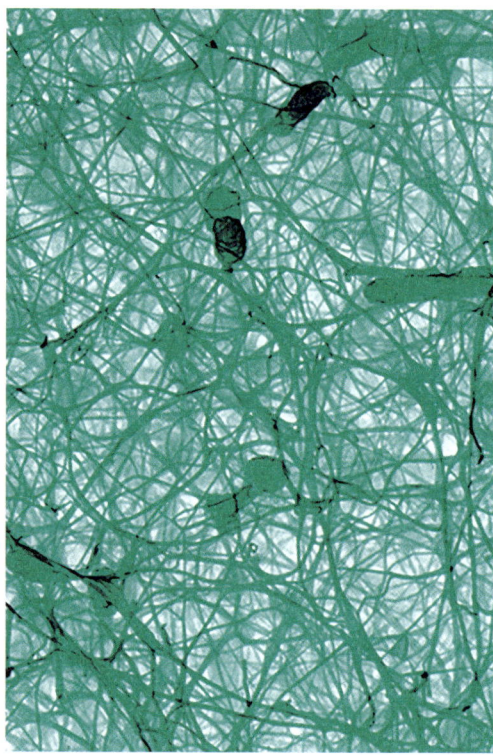

Fig. 5
The world of fibers—
shared agency.
Left panel: *Space Loom #1* by
Hella Jongerius (Installation
at Lafayette Anticipations,
Paris, 2019).
Right panel: Cells of traditional
vinegar Acetobacter bacteria
entangled in a network of
self-produced cellulose fibrils.

Might the idea of the extended organism not also be applied to human-made structures—to our cities, which channel the flow of people and materials along with energy, information, and capital—to the advantage of social coexistence? Could our built environment also be understood as an externalized collective *physis*? And what would an architecture look like that, instead of being built from bricks or stones, reflected the principles of spinning and weaving, thereby uniting technical needs with a physiology that transcends both the individual and the human species? Perhaps this would be an adaptive, active architecture, resembling the structures explored by Hella Jongerius in her experiments with three-dimensional weaving (Fig. 5). In her artistic work, the industrial designer has critically questioned the use of materials and production systems. The

4   See J. S. Turner, *The Extended Organism: The Physiology of Animal-Built Structures* (Cambridge, MA, and London: Harvard University Press, 2000), 7. In this context, Turner also refers to "external physiology" (ibid.).

weaving and the spinning of the threads are in the foreground in their technical as well as in their social and cultural or even spiritual dimension. The ancient metaphor of the cosmic loom serves as her inspiration for the search for a new relationship with the objects we produce, in order to arrive at a better, more sustainable relationship with this planet.[5] She also translates this into concrete research, such as with her self-designed 3D loom, which is operated collectively, leaves room for the individual and his/her creativity, and produces three-dimensional objects (Fig. 4) that can have very different functions depending on the material and size. In her exhibition *Hella Jongerius: Woven Cosmos* at the Martin-Gropius-Bau in Berlin in 2021, she presented 3D structures woven from strips of solar cells that could transform solar energy into electricity or, consequently, into autonomous movement. The potential of these weavings lies in their economical use of materials, their elastic stability, and their modularity. Jongerius says about these prototypes: "Textile is both the strongest and lightest of constructs: we are engineering folded constructions by creating multiaxial weaving, which have embedded power structures. In other words, we are making modular structures as pliable architecture."[6] Such active materials not only aim at innovative technical solutions for architecture, but far more fundamentally bring to the fore the relationship of the creating individual (the maker) to the materials used and the objects created, in order to arrive at an awareness of the entanglement of the human being in the world (Fig. 5), i.e. the "shared agency" already mentioned.

THE "TEXTILITY OF MAKING" AND ITS ROLE IN A NEW "ECOLOGY OF LIFE"

Developing awareness of a material's texture, its material quality, requires active senses and an active intellect. This comes naturally to the person who makes. Manual production inevitably involves an implicit and explicit awareness of the materiality and relational character of shaping. This is what the British anthropologist Tim Ingold calls the "textility of making," which he discusses with the example of weaving.[7] According to Ingold, what is at stake here is not craft per se but a different "ecology of life."[8] We think of our products and creations solely in terms of their rational and economic construction instead of relating them to our collective and individual existence, which cannot be understood without being embedded in environments.

Although organic life has a blueprint in the form of DNA, it emerges, exists, and grows only through its interactions with other factors, such as other living beings or "dead" matter. Organism and environment cannot be strictly separated, with activity being shared and distributed because both can exist only in an interdependent process with a generally open outcome. The organism can orchestrate the flows of material and energy that pass through it and keep it alive by "weaving," "entwining," and "felting." In doing so, however, the organism changes what surrounds it and has to cope with the environment's reactions and transformations. It can be argued,

---

5   See H. Jongerius and S. Rosenthal, "Many Hands, Many Minds. A Conversation between Hella Jongerius and Stephanie Rosenthal," in *Hella Jongerius: Woven Cosmos*, exhibition at Martin Gropius Bau in Berlin (April 29–August 15, 2021), booklet, ed. S. Rosenthal and C. Meister (Berlin: Gropius Bau, 2021), 20–29, here 27: "We've ruined the planet but we can design our way out of this mess. In that sense, I see the cosmic loom as a metaphor for weaving a new texture for the world, of navigating fate by choosing your yarn in our Anthropocene world."
6   Ibid., 29.
7   See T. Ingold, "The Textility of Making," *Cambridge Journal of Economics* 34 (2010): 91–102, here 92–93.
8   See the chapter "Culture, Nature, and Environment: Steps to an Ecology of Life," in T. Ingold, *The Perception of the Environment: Essays on Livelihood, Dwelling and Skill* (London and New York: Routledge, 2000), 13–26.

therefore, that the recognition and practical experience of the activity of the material around us is a prerequisite for our beneficial re-connection and much-needed reintegration into our planet's active material cycles.

A first version of this article was included in the now out-of-print booklet to the exhibition *Hella Jongerius: Woven Cosmos* Martin Gropius Bau, Berlin. The exhibition was on view from April 29 to August 15, 2021. Weaving has always had a special importance for the Dutch artist and designer Hella Jongerius (b. 1963), whose work attempts to reconcile the history of this cultural technique with ideas for future possibilities, such as three-dimensional weaving or woven architecture. The exhibition presented a selection of Jongerius's artistic works that addressed both the traditional and the new, speculative possibilities of spinning and weaving. At the center of the exhibition was a functioning workshop in which visitors could participate — along with the artist and her assistants — in processes of spinning and weaving. The works on display and those produced in the course of the exhibition were contextualized by questions about the relation between industrialization and craft, production and sustainability, material and technique. In addition, this article pointed out the striking structural and functional similarities between the entangling and weaving of fibers in human culture and biological systems across the scales. For more information on the exhibition, see: https://www.berlinerfestspiele. de/en/gropiusbau/programm/2021/ hella-jongerius/ausstellungsguide.html (accessed 26.10.2021). See also the longer and more detailed version of the booklet article in the online journal of the Gropius Bau at https://www.berliner-festspiele.de/en/gropiusbau/programm/ journal/2021/regine-hengge-karin-krauthausen-the-event-of-a-fibre.html (accessed 26.10.2021).

The authors would like to acknowledge the support of the Cluster of Excellence *Matters of Activity. Image Space Material* funded by the Deutsche Forschungsgemeinschaft (DFG, German Research Foundation) under Germany's Excellence Strategy — Exe 2025 — 390648296.

Bacterial Loom

Team
Dr. Bastian Beyer, Dr. Skander Hathroubi, Prof. Dr. Regine Hengge
Faculty of Biology — Department of Microbiology, Humboldt-Universität zu Berlin

Context
Research project, Cluster of Excellence *Matters of Activity. Image Space Material*,
Humboldt-Universität zu Berlin, 2020–22

Material
Bacterial cellulose, textile thread, nutrients

→ Lifting process of bacterial cellulose.

↖ Continuous "harvest" of bacterial cellulose. Each layer takes 5 days to grow.

While conventional cellulose extraction from plant materials is an energy-intensive and chemically laden process, growing bacterial cellulose in the form of biofilms offers an alternative source of highly pure cellulose, without the need for further extraction processes. The interdisciplinary research project *Bacterial Loom* experiments with new strategies of making textile materials, by working with the distinct growth and materiality of bacterial cellulose. *Bacterial Loom* explores a setup of co-weaving with bacteria in which two textile systems, bacterial cellulose and human-made textile scaffolds, intersect and interact.

Bacterial cellulose has outstanding properties such as high tensile strength, high water absorption capability and biocompatibility. Specific types of bacteria are capable of growing bacterial cellulose in the form of complex micro structures called biofilms. The bacterium *Komagataeibacter hansenii* produces a biofilm that forms at the interface between air and a liquid nutrient medium, consisting mainly of nanocellulose. Although the biofilm appears as gel-like on a macro-scale, it is inherently fibrous on a micro-scale.

In *Bacterial Loom* the cellulosic biofilm forms around an array of vertical threads. Once the biofilm has grown, the threads are simultaneously lifted, allowing the bacterial cellulose to dry suspended in the textile scaffold while another biofilm is formed below from the same culture. The points where the biofilm and threads intersect are not only structurally relevant but are highly active areas during the growth process. The capillarity of the thread increases the surface area of the air-medium interface, leading to areas with denser bacterial cellulose.

Analogous to the practice of weaving, *Bacterial Loom* lets two textile systems intersect in a perpendicular manner. While the vertical "warp" yarns are introduced through human interaction, the "weft" layer is established by the growth of the biofilm. This process of *co-weaving* allows for two inherently different textile materials to merge into one structure.

While the yarn setup remains the same over the course of the experiment, the individual biofilms vary significantly between each layer. This variation is due to the differences between the individual biofilms, their growth, microstructure, and drying behavior as well as the depletion of nutrients after each growth cycle. The experimental setup allows us to observe individual bacteria cultures and their adaptation to changes in their physiochemical environment over longer periods of time. Further, it allows for the study of how the complex cellulose structure of the biofilm is affected by mutation of the bacteria as well as changes to the nutrient composition.

BACTERIAL LOOM

↑ Bioreactor setup (with comparative study of two bacteria strains).

← Detail of dried bacterial cellulose.

Aerial Construction

Team

Ammar Mirjan, Prof. Fabio Gramazio, Prof. Matthias Kohler, Gramazio Kohler Research, ETH Zurich
Federico Augugliaro, Prof. Dr. Raffaello D'Andrea, IDSC, Institute for Dynamic Systems and Control, ETH Zurich

Context

Research prototype, bridge structure made through aerial robotic construction, 2013–15

Material

Quadrocopters, ultra-high-molecular weight polyethylene rope (Dyneema), custom rope dispensers, scaffolding, sensing-motion capture system

The project *Aerial Construction*, a collaboration between Gramazio Kohler Research and the Institute for Dynamic Systems and Control at ETH Zurich, investigated the architectural potential of flying robots. Because aerial robots are kinematically decoupled from a conventional ground-based working environment, they offer distinctly new forms of construction compared to conventional construction devices. First, the working space range of aerial robots is not limited, making them capable of operating at full architectural scale. Second, their ability to move autonomously through and around existing structures makes them ideal for intertwining elements. When these abilities are combined with control algorithms that enable the aerial robots to cooperate and perform construction tasks simultaneously in a group, entirely new building forms become possible.

To leverage the unique skill of a flying robot to steering and aggregating material in three-dimensional space, the project identified the spatial interweaving of tensile rope structures as an appropriate construction technique. For an algorithmic controlled aerial choreography, information is sent to the flying machines via a customized wireless infrastructure. The flying robots move independently of the structure they are building before they fasten the rope at a desired location in space. The machines can maneuver around and in-between structures to pilot rope around objects and tie knots.

The project brings forward the making of textile structures as a concatenation of basic computationally programmed building operations, so-called primitives, to design a complex structure. The building primitives that were identified as being essential for the proposed construction system are: the knot (for creating tensile joints), the link (for spanning space by linking two knots with rope), and the braid (for interworking tensile

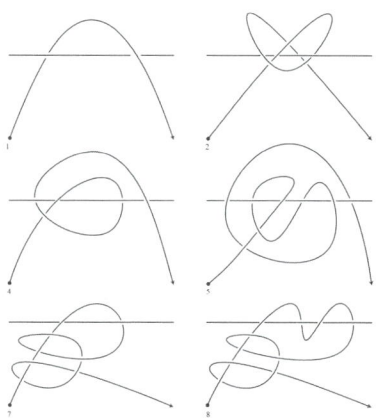

↑ Planar projection of examples of knots. The different knots constitute repeatable basic crossing primitives around the support bar and the standing part of the rope.

↗ Exemplary drawing and notation of half-hitch with an additional round-turn. The knot starts with a round-turn on the bar (1). Then a hanging segment is created on the standing part of the rope (2–4), through which the vehicle pulls the rope (5). Afterwards, the overhand knot is tightened by creating another turn around the support element (6–8).

members). By combining these elements, various tensile structures can be computationally described and physically realized.

The utilized Dyneema rope distinguishes itself for aerial construction due to its high strength and low weight. Its weight-to-strength ratio is around 8–15 times lower (better) than that of steel. The resulting full-scale, load-bearing footbridge, spanning 7.4 m, is built autonomously by the flying machines. The experiments were performed in the Flying Machine Arena, an indoor space for aerial robotic research at ETH Zurich.

All images: © Institute for Dynamic Systems and Control and Gramazio Kohler Research at ETH Zurich.

↑ Simultaneous erection of the three main links by three vehicles. Two machines start building the handrails on one side, while the third machine fixes the footrope on the other.

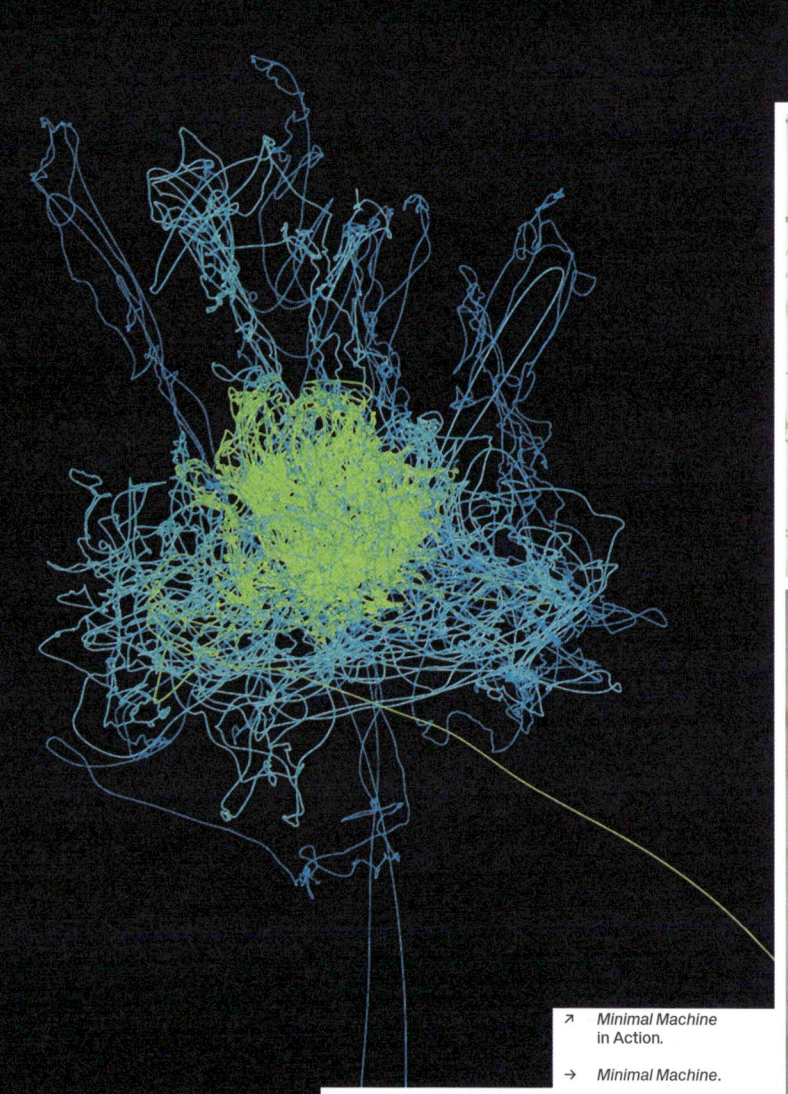

Augmented Spinning

Team

Elaine Bonavia, Prof. Dr.-Ing. Karola Dierichs
Supervision *Minimal Machines*: Prof. Dr.-Ing. Karola Dierichs, Jessica Farmer,
Dr. Mareike Stoll—Weißensee School of Art and Design Berlin / Cluster of Excellence
Matters of Activity. Image Space Material, Humboldt-Universität zu Berlin, Elaine Bonavia,
Guest Prof. Mika Satomi—Weißensee School of Art and Design Berlin, Dr. Laurence Douny,
Dr. Michaela Eder, Nikolai Rosenthal, Charlett Wenig—Max Planck Institute of Colloids and
Interfaces / Cluster of Excellence *Matters of Activity. Image Space Material*, Humboldt-
Universität zu Berlin, Students: Madleen Albrecht, Sara Hassoune, Elisa Martignoni,
Sebastián Plaza Kutzbach, Marie Rasper, Weißensee School of Art and Design Berlin

Context

Design Research as part of *Minimal Machines I*, MoA Design Research Studio,
Weißensee School of Art and Design Berlin, 2021

Material

Nylon (Polyamide) rope 2 mm, hemp rope 3 mm, processed with Inside-Out Body Tracker,
Electronic Bluetooth Module, vibration motors, glove, 3D NURBS modeling software

AUGMENTED SPINNING

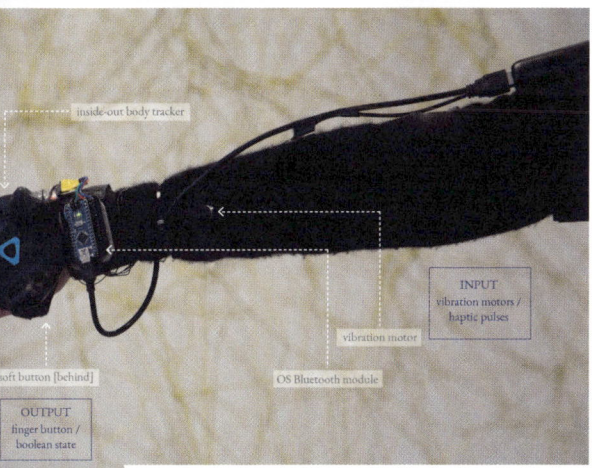

inside-out body tracker

INPUT
vibration motors /
haptic pulses

vibration motor

soft button [behind]   OS Bluetooth module

OUTPUT
finger button /
boolean state

The project *Augmented Spinning* investigates
the potential of haptic technologies in the
process of constructing architectural material
systems at scale. Traditionally, spinning entails
a careful negotiation between bodily rhythm
and the spinning machine. This project extends
the intimate act of spinning into the spatial to
explore the possibilities of making architecture
from the place of embodied movement and
augmented reality (AR). It thus contributes to
ongoing research on the integration of human
cognitive intelligence into digital fabrication
processes that is, among others, explored by
Atanasova and collaborators.[1]

Avram defines AR as the overlay of the human
sensory system with virtually generated
information in a real-time feedback loop.[2]
Augmented Spinning on the one hand allows
the collection of information about the construction procedure, and on the other it processes this information and returns it itera-
tively into that very same process. Together,
these processes begin to derive a collective
construction intelligence between the maker,
the material, the augmentation device and the
platform.

1   L. Atanasova, D.Mitterberger, T. Sandy, F. Gramazio, M. Kohler, and K. Dörfler, (2020)"Prototype as artefact" in *ACADIA
     2020: Distributed Proximities / Volume I: Technical Papers [Proceedings of the 40th Annual Conference of the Association
     of Computer Aided Design in Architecture (ACADIA)]*, 24–30 October 2020, ed. B. Slocum, V. Ago, S. Doyle, A. Marcus, M.
     Yablonina, and M. del Campo, 516-25.

2   H. Avram, "Augmented Reality," in *Oxford Encyclopaedia of Aesthetics*, Second edition, Vol. 1, ed. M. Kelly (Oxford: Oxford
     University Press, 2014), 232–36, https://doi.org/10.1093/acref/9780199747108.001.0001.

The project was initiated as part of the MoA Design Research Studio *Minimal Machines I*, and focused on testing an augmentation device in the form of a glove. This was prototyped as a wearable machine and envisaged as a tool that equips the maker with the minimal amount of technology needed in the process of making.

The machines were to be used as devices in conjunction with a designed material. Given that one of the core paradigms of designing matter is the abolition of machines in favor of matter's inherent activity, these machines were to be minimal.

The augmented glove was able to provide haptic feedback throughout the construction process by means of a number of vibration motors placed on the hand and arm. Addition-ally, an inside-out body tracker coupled with an open-source electronic Bluetooth module makes it possible to send and receive posi-tional and vibration signals between the simu-lation environment and the physical space. The augmented glove is also capable of recording spun point positions through an integrated button, not only allowing for the creation of a digital syntax but also providing insight and visibility to the different forms of movement and the layering of bodily rhythms embedded in the process.

Finally, the Augmented Spinner receives signals from the haptic motors that act as an indicative compass, implying a direction based on the status of the structure in progress. This starts to facilitate a feedback loop and therefore also a co-creation process. As Donna Haraway

discusses in her *Cyborg Manifesto*, the blurring of the line between human beings and machines causes us to examine human processes more closely, and to consider cyber-physical systems in architecture as inherently embedded systems.[3]

D. Haraway, "A Manifesto for Cyborgs: Science, Technology, and Socialist Feminism in the 1980s," in *Feminist Social Thought: A Reader*, ed. D. Tietjens Meyers (New York, London: Routledge, Taylor & Francis Group, 1997), 502–31, https://doi.org/10.4324/9780203705841-39.

Slip-Form Rock Jamming, Modular and Morphable Rock Jamming

Team

Slip-Form Rock Jamming:
Bjorn Sparrman, Schendy Kernizan, Jared Laucks, Prof. Skylar Tibbits, Self-Assembly Lab, Massachusetts Institute of Technology MIT, Google

Modular and Morphable Rock Jamming:
Zach Cohen, Nathaniel Elberfeld, Andrew Moorman, Schendy Kernizan, Jared Laucks, Prof. Skylar Tibbits, Self-Assembly Lab, Massachusetts Institute of Technology MIT
Prof. Douglas Holmes, Boston University

Context

Collaborative research projects, 2017–19

Material

Fibers, strings, loose rocks and gravel, construction by slip-form molding, falling-string deposition method, superjamming

↑ The wall-slab was formed by fabricating a vertical super-jammed wall, rotating into the horizontal.

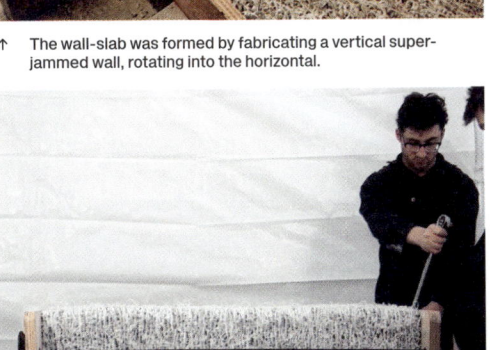

↑ The "beam-arch" demonstrated the morphability of super-jammed structure.

Jammed structures comprise a solid disordered mass of granular particles capable of compacting under their own weight. These aggregated structures of gravel hold the potential of self interlocking and achieving high strength and load-bearing capabilities. The aggregated mass transits between a loose state when it is in flow, and a solid state when it is compacted. With the introduction of layered fibers and threads the friction between the single aggregates can be amplified to make these granular structures stable and solid.

In the two projects *Slip-Form Rock Jamming and Modular and Morphable Rock Jamming*, MIT researchers use loose fibers and threads that are placed in layers between the gravel to increase the friction and enable the construction of loose-fiber reinforced granular columns, slabs and walls. Local fibrous material and simple gravel can be used and processed into a variety of different structures. For the project *Slip-Form Rock Jamming* a zig-zag wall was created by alternately filling gravel and coconut husks into a slip-cast mold.

For the project *Modular and Morphable Rock Jamming* the technique of "superjamming" was applied to achieve horizontal building elements. A "falling-string deposition" method introduces layers of looped string as reinforcement while filling loose rock into a vertical slip-mold form. By post-tensioning implemented threaded rods fastened to wooden end plates the granular slab is compacted and undergoes a phase-transition into a solid-like state. When rotated horizontally it can withstand significant loads and even become morphable into arches. Different types of rock (red slate, gravel, crushed marble, and gneiss) and three types of yarn (twisted nylon, polypropylene twine, and acrylic) were tested for their individual properties affecting the stability of the jammed structures. Rocks with lower sphericity (e.g. red slate) produced structures of a higher resilience. The advantage of using this dry assembly compared to a matrix-bound composite material is in the possibility to easily disassemble the system and make way for transformable and reversible structures. The coiling thread simplifies significantly both the construction and the disassembly simply by pulling a string.

← String was deposited, while rocks were poured by hand. Falling string deposition method with rocks poured by hand.

↑ Coconut husks and loose gravel were used to fill a slip-form mold to continuously build a load-bearing zig-zag wall.

Woven Paper Bridge

Team

Dr. Lorenzo Guiducci, Maxie Schneider, Josephine Shone, Prof. Christiane Sauer, Cluster of Excellence *Matters of Activity. Image Space Material*, Humboldt-Universität zu Berlin

Context

Research project and full-scale prototype, Cluster of Excellence *Matters of Activity. Image Space Material*, 2021

Based on *Sticky Stacks* design project by Eva Eckert, Serafina Baucken, and Josephine Shone (see p. 77), MoA Design Research Studio *Scaling Nature (2): Fibers, Muscles & Bones*, Weißensee School of Art and Design Berlin, 2020

Material

Copy paper sheets 80g/m², 10.5 × 29.7 cm

Processed with manual interleaving, paper drill, paper cutter

→ Tensile tests on two interleaved paper stacks made with 50 or 100 sheets. The geometric amplification of friction shows a highly non-linear increase in maximum load before slipping.

↓ Close-up of interleaved paper stacks.

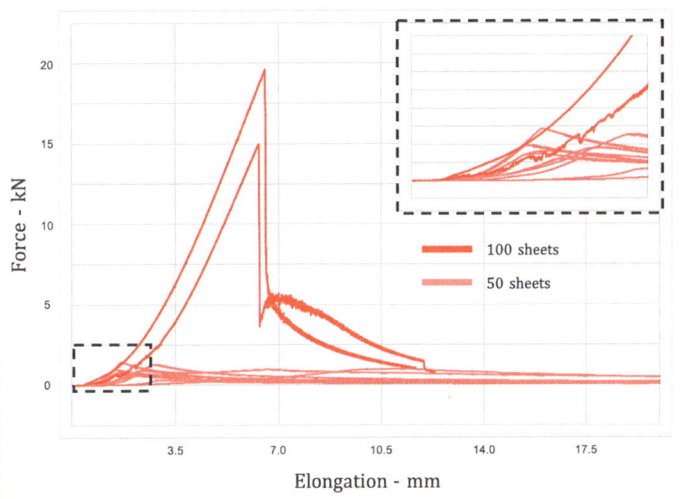

Woven Paper Bridge showcases an experimental construction that is inspired by muscle contraction. Muscle fibers contain bundles of interleaved protein filaments that when moving past each other generate tiny contraction forces. These are dramatically amplified by the number of protein units arranged in parallel. Mimicking muscle-tissue structure, two paper stacks were fixed and interleaved (see p. 76). Upon pulling the two stacks apart, a perpendicular force is created that holds the sheets together at the overlap region. As a result, the total friction between the sheets grows exponentially—a phenomenon known as "phone book enigma."

Paper is mainly made of cellulose, the most abundant biopolymer on earth, and is characterized by a high-tensile modulus due to the specific orientation of cellulose fibers. Tensile tests showed that two stacks made with 100, 80g/m² sheets of standard white copy paper, and interleaved with an overlap of nine centimeters, withhold more than one ton before the connection begins to slip.

Adopting aspects of woven textiles, we developed a bottom-up process to assemble cut paper sheets into a meter-sized bridge. In this way, the benefits of paper (light, strong,

Due to their low weight and excellent mechanical properties, new applications for paper-based materials are increasingly being investigated in architectural research, which, however, tend to rely on joining techniques such as gluing, pinning, or folding. In this regard, the woven paper bridge proposes a sustainable and temporary system, based on reversible, friction-based connections that can even be implemented as a DIY principle.

Conventionally, tons of waste paper are recycled in an energy and resource-intensive process and mostly downcycled to short-lived products such as packaging or newsprint. The project aims at the direct reuse of waste-paper sheets after collection from the local recycling infrastructure. Paper only needs to be sorted, cut, and joined, a process that can be implemented on a local scale.

WOVEN PAPER BRIDGE

paper stack

sorting
cutting
interleafing

material
system

waste paper

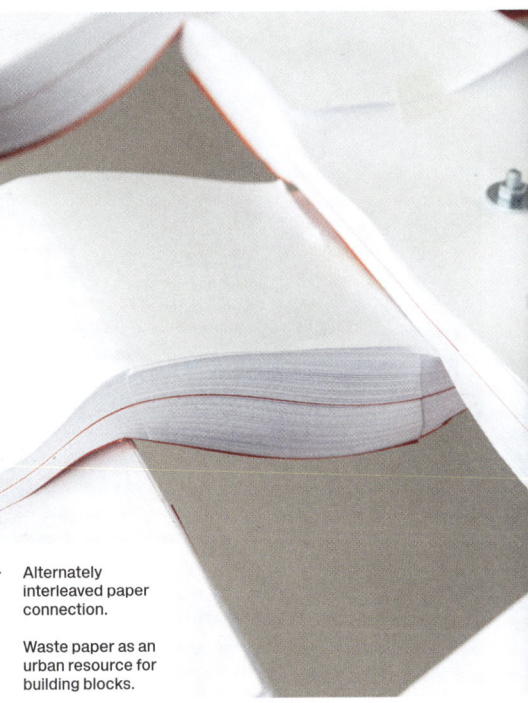

→ Alternately interleaved paper connection.

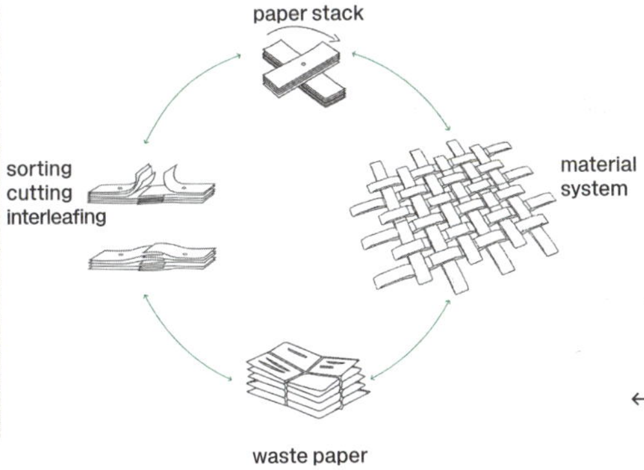

← Waste paper as an urban resource for building blocks.

recyclable) were augmented via a geometric design that improves some of paper's known issues: stacking the sheets increases resistance to crack formation, interleaving provides high friction and hence large tensional loads, cross-weaving allows for better load distribution. With a length of 4.5 meters and a width of one meter, the bridge contains 240 stacks with a total weight of fifty-five kilograms. Assembled as a plain weave, the bridge can carry up to five people (400 kilograms).

Remarkably, similar strategies have occurred across different cultures: following a centuries' old Inca ritual, natives of the Cusco region in Peru build Q'eswachaka, a woven-rope bridge carefully crafted from locally harvested grass.[1] As an urban interpretation, waste paper can be used as a valuable resource, thus indicating a more sustainable approach to construction that calls for more care, and leverages locally available resources.

1   E. Muscio and J. Anaya Díaz, "Braided rope with vegetable fibers for the construction of the Inca bridge of Q'eswachaka (Peru)." In *Building Knowledge, Constructing Histories* (Boca Raton: CRC Press Balkema, 2018), 977–81.

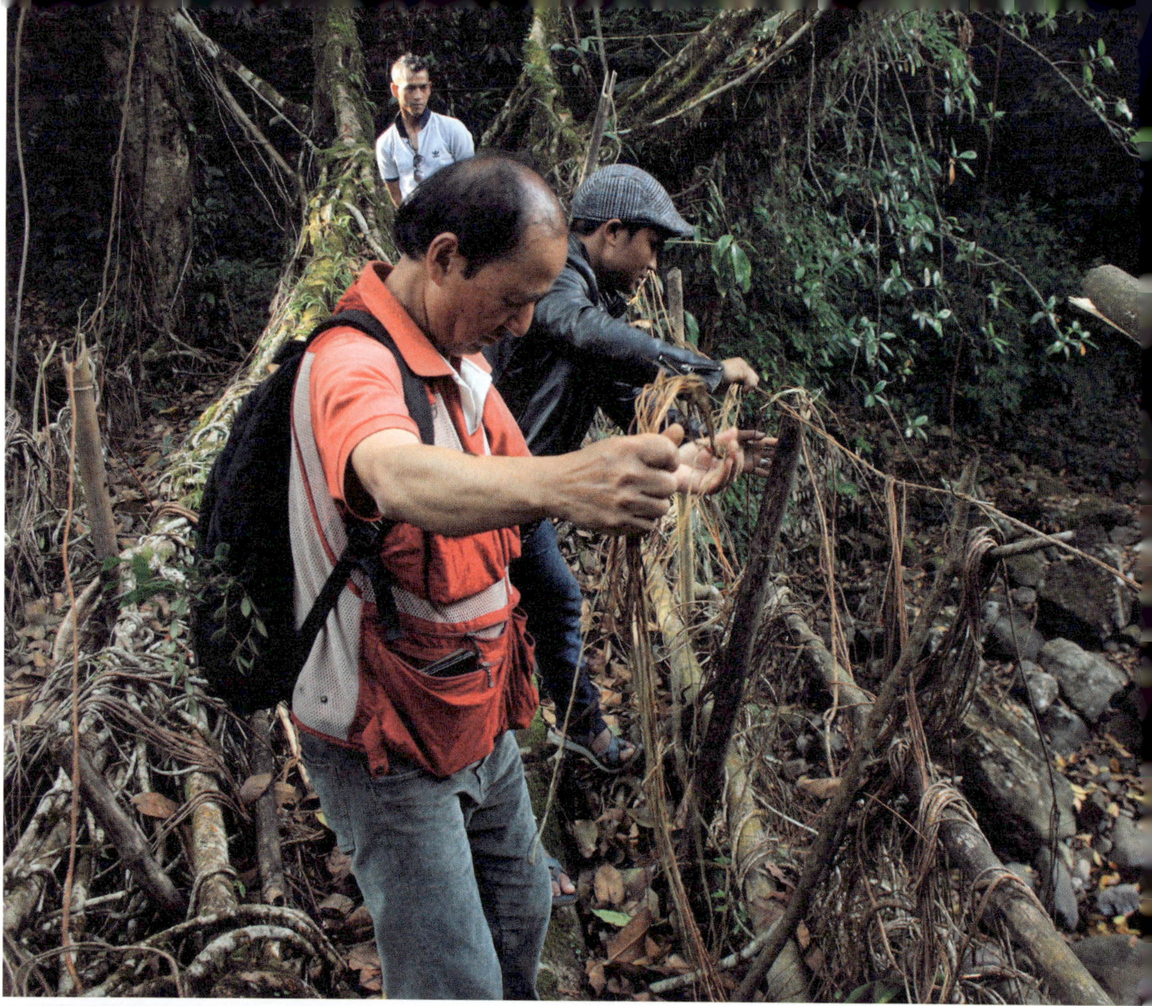
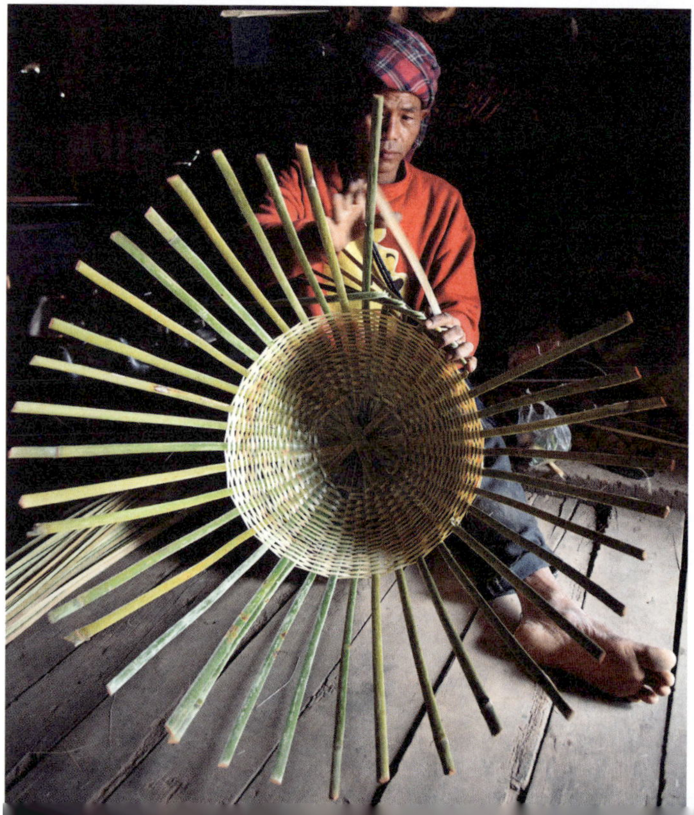
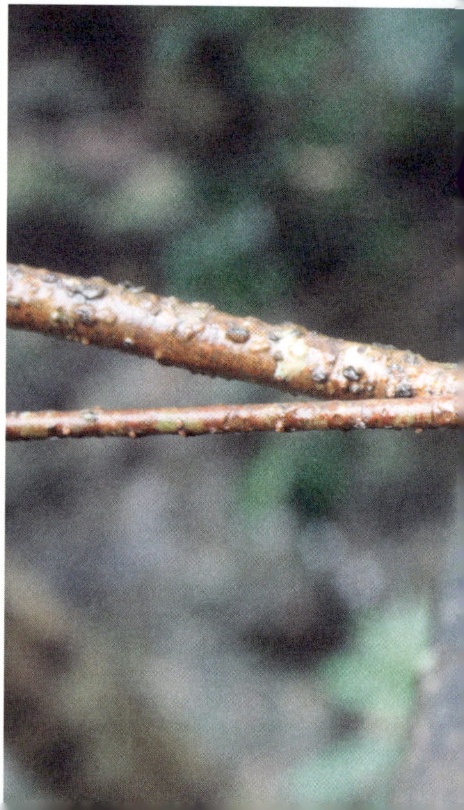

Living Root Bridges (Jingkieng Jri)

Team

The Garo, Khasi, and Jaintia Tribes, Meghalaya, India

Context

Weaving heritage and building practice, ca. 1500 to present
Researched by Wilfrid Middleton, Prof. Dr. Ferdinand Ludwig, Green Technologies in
Landscape Architecture, Technical University of Munich

Material

Ficus elastica roots, bamboo, cultivated by planting, guiding and manual weaving

LIVING ROOT BRIDGES

Traditions of weaving are an integral and cele-
brated part of the cultural heritage of Meghalaya, North-East India. Clothes are woven from
Meghalaya's natural silk, in locally specific
patterns[1] and using techniques and tools as
diverse as the state's many linguistic dialects.
Bamboo and rattan weaving for stools, mats,
hunting and fishing tools, and baskets are well
recognized as weaving cultures and used in
day-to-day life. Weaves find many uses in the
traditional houses of the state's three main
tribes (Garo, Khasi, and Jaintia), in which
bamboo is used in dense and sparse lattices, as
well as whole stems and thin slivers to utilize the
natural mechanical properties of the plant.

Living root bridges, or *Jingkieng Jri*, are grown
by guiding, twining, and weaving the living roots
of the Indian Rubber Fig (in Latin *Ficus elastica*,
in Khasi *Dieng Jri*) across rivers and canyons.
More than seventy bridges link villages to their
farmland, markets, and the wider state infrastructure, providing a vital rural transport
network. The process of growing a living bridge

1   A. Karolia and B. Ladia, "Traditional textiles and costumes of Karbi and Biate tribes of Meghalaya," *Indian Journal of
Traditional Knowledge* 11, 2 (2012).

LIVING ROOT BRIDGES

begins with growing a *Ficus* tree on a riverbank. A bamboo bridge provides a temporary crossing and a scaffold onto which the aerial roots of the tree are guided. The roots are thin and flexible, growing from the branches and trunks in search of water and nutrition. They are known to grow fifty metres in length before attaching to soil or rock, making them ideal for spanning wide rivers and canyons.[2] The bridge becomes a dense network of young roots wrapped around bamboo. Once they find ground, the roots grow tight and thicken, creating rope-like tension elements between the water source and the tree. Over time, as the root continues to thicken, this tension is released, and the root can provide a strut on which the tree can lean. These then produce supple "daughter" roots, which are then woven into the bridge, eventually growing thicker and stronger themselves.

The first supple roots that are guided over the bamboo structure thicken and become central roots to support the entire structure, replacing the bamboo as it decays. Then, just as a woven cloth uses a thicker thread for the warp than the weft, the thin "daughter" and "granddaughter" roots are woven between the larger compressive and tensile roots. Flexible roots are knotted, forming a lattice. During the tension phase of growth, these knots are pressed together, increasing the chance of a successful graft. As the lattice ages and more roots are woven, it widens the bridge, forming the deck and handrails. The structural network is made up of multiple paths between diverse elements,[3] improving earthquake resistance.

In contrast to pre-planned clothing and bamboo architecture with set patterns, the form of the living root bridge is developed over time. Builders react to unpredictable aspects like individual root positions and growth rate, the bridge's use, and weather, building their understanding of *F. Elastica's* underlying growth characteristics and local conditions. In turn, the tree's growth is a response to the builders' work, creating a conversation from which each bridge uniquely emerges.

2   F. Ludwig, W. Middleton, F. Gallenmüller, P. Rogers, and T. Speck, "Living bridges using aerial roots of Ficus elastica—an interdisciplinary perspective," *Scientific Reports* 9 (2019): 1–11, https://doi.org/10.1038/s41598-019-48652-w.

3   W. Middleton, Q. Shu, and F. Ludwig, "Representing living architecture through skeleton reconstruction from point clouds," *Scientific Reports* 12 (2022): 1–13.

Dorze Architecture

Team
> Dorze People, Arba Minch Area, Ethiopia

Context
> Woven Bamboo Houses, traditional and contemporary building practice, described by Dr. Petra Gruber, Ethiopian Institute of Architecture, Building Construction and City Development EiABC at Addis Ababa University, Ethiopia, 2014

Material
> Bamboo, machete, knife, other hand tools, hand weaving

←← Dorze house in the building stage. Bamboo material is prepared and dried on the site.

← Interior view with temporary scaffold structure.

↓ Details of patched cladding with aged and new repair material.

In traditional architecture, applied technologies are strongly connected to cultural practices of creating agricultural and domestic products. Weaving is used by people around the world to create not only textiles, but also furniture and wall elements. Especially in Asia where bamboo is abundant, separation walls made of split and woven bamboo are very common, but also in other areas where bamboo is grown, comparable practices have evolved. In Ethiopia, East Africa, two outstanding architectural typologies exist that are constructed of woven bamboo. The Sidama people in South Ethiopia build onion-shaped houses with a central pillar. This case study describes the dome-shaped houses of the Dorze people in the area around Arba Minch, which are constructed entirely with bamboo.[1]

1   P. Gruber and K. Datta, "Construction Aspects in Ethiopia's Architectural Traditions: A Comparative View," *Journal of Traditional Building, Architecture and Urbanism*, 2-2021, 11/2021.

In this region of South Ethiopia, the climate is warm and humid, vegetation is lush, but there is little primary forest, so wood is scarce and expensive. The main crop in the Dorze region is "enset," false banana, or *Ensete ventricosum*, a plant similar to the banana, but without edible fruits. The plant is very easy to grow and highly valued since all parts of it are used for the production of food (a processed pulp from the stalk and rhizome), fibers for textiles and also as an alternative to bamboo leaves for the thatch of the houses. The Dorze people construct their houses out of split bamboo stems that are flattened and dried.[2] The dome-shaped buildings are essentially architectural-sized woven baskets that are stuck into the ground. An annex part is added to the main volume and serves as an entrance. Diameter and height are around eight meters, so horizontal bamboo sticks are used as temporary scaffolds at the building stage.

The houses are made by an expert builder, who may have help from the owners. The production time is two to three weeks, not including material preparation. The finalized constructions are then clad with false banana or bamboo leaves, depending on the region. The detailing of the tip is important because the roof tips are in general very vulnerable to water entry and require protection from the rain. The interior space is also divided by woven mats. As the houses age, the thatch is patched and replaced. As humidity slowly decays the bamboo sticks connecting to the ground, they rot away, and the houses become slightly smaller over time. These extremely light constructions can be moved to other locations and can last up to fifty years. The Dorze villages consist of clusters of compounds that are fenced off against public walkways with woven-bamboo walls. The Dorze people are also famous for their textile-weaving techniques and fabric production.

2   J. Olmstead, "The Dorze House: A Bamboo Basket," *Journal of Ethiopian Studies* 10, no. 2 (Institute of Ethiopian Studies, 1972): 27–36.

Acknowledgements

Petra Gruber documented the Dorze woven architecture in 2014 during her research stay in Ethiopia. Thanks go to: Ethiopian Institute of Architecture, Building Construction and City Development EiABC at Addis Ababa University, Ethiopia.

DORZE ARCHITECTURE

←← Dorze house, with enset plants (*Ensete ventricosum*) in the foreground, Ethiopia 2014.

← The dome with an annex being woven into the main structure by the builder. The horizontal sticks are a temporary scaffold during the weaving process.

Topologies of Weaving—A conversation on Mathematics and Weaving with Myfanwy Evans and Alison Grace Martin

EBBA FRANSÉN WALDHÖR:
Myfanwy, as a mathematician and Professor for Applied Geometry and Topology at the University of Potsdam, and Alison, as an artist and researcher based in Tuscany, you're both working in weaving in contexts not usually associated with textile practice. What led you to weaving?

ALISON GRACE MARTIN:
When I originally started weaving nearly thirty years ago, I was driven by practicalities. I was trying to live in a self-sufficient, sustainable way and I needed to make things for my garden—very rudimentary types of structures. Because of a shortage of cash, I used material that grew on the land, which was bamboo. I was looking at images of hand-made objects and basket-weaving in other continents for inspiration. Quite quickly it became clear to me that weaving is related to geometry, and I started making more detailed studies of how geometry can help to build things that are strong and efficient. It turns out that the ancient Japanese tri-axial weaving pattern *kagome*, for instance, is a good way to form shapes. Introducing pentagons to the two-dimensional weave allows it to bend and become a three-dimensional object. I began to weave experimental shapes of negative curvature simply out of curiosity.

MYFANWY EVANS:
My starting point was the spectacular surfaces and structures that are prevalent in biological cells. These curious mathematical objects were described in the nineteenth century. I started out on the thought experiment of whether proteins and polymers crystallize into surfaces to make entangled 3D patterns in space. Exploring this with experimental mathematics, I visualized how lines or curves on negatively curved surfaces would look as a 3D structure, without the surface. Imagine I have a membrane, like a soap film, with long proteins crystallizing onto that surface. Then my soap film bursts, the membrane disappears and I'm left with a tangle of three-dimensional lines in space. I can then ask the questions: What do these entangled structures look like? What can I say about them topologically?

CHRISTIANE SAUER:
Topology is a central term in your work. How do you define it?

ME:
Let me give an example of the difference between geometry and topology, to clarify what we mean by topology. Topology is a discipline of mathematics, about equivalences of structures. Imagine I have a small square of knitted fabric that has a particular geometry—I can rotate it around without changing the geometry. That would be a geometric equivalence. If I take my piece of fabric and stretch it, however, I change its geometry—it now has a different size and shape. My stretched structure and the original structure are geometrically different but topologically equivalent under this particular deformation. The stretching of the knitted structure can be thought of as an ambient isotopy—which is one kind of topological equivalence. If you deform the knitting such that the yarn can pass through itself, you're left with just a single yarn. This deformation is a homomorphism, another type of topological equivalence, and gives information about what we're tangling. You have this whole hierarchy of things that you can describe as equivalent. That's really what topology is—looking at what's equivalent under very specific deformations or changes.

TOPOLOGIES OF WEAVING

CS:
Topology seems related to parametric design: not defining one specific shape, but the rules for forming it. The design outcomes are possible variations that can be generated with the given parameters. When programming became available as a design tool, it shifted the role of the designer from being the creator of unique forms to being an explorer of formal possibilities: not saying "This is the design I want", but "This is the way I want to design."
I also find this question of whether geometry is something we find or something we invent really interesting. Geometry is both embedded in culture—like the traditional basket-weaving crafts—and in biological material structures found in nature. How far is your work concerned with discovering the inner logic of things, and how much of it is about creating something new?

AGM:
My work isn't creative in the sense that it's my imagination producing free-forms shapes. The shapes that inspire me are fundamental, natural structures—from crystallography, chemistry, structural biology. They have a lot of advantages: they're very strong—and light. It seems like a natural opportunity to copy these rather than invent something new that doesn't exist. I think it's also interesting to look at other natural weavers, like insects and birds, and the patterns they've come up with. Nature is interesting in two ways: One is the actual structure of the organisms that

→ Hand-tool for splitting bamboo canes to make them pliable for weaving.

↓ Woven version of a minimal surface. Division of space into two congruent regions.

↘ Eight cubic cells of the Schwarz P (primitive) surface. A woven study like an infinite periodic minimal surface, related to a hypothetical fullerene and graphitic structure. The weaving pattern is made by interlacing straight-cut paper strips, which wrap around the surface following approximately geodesic paths.

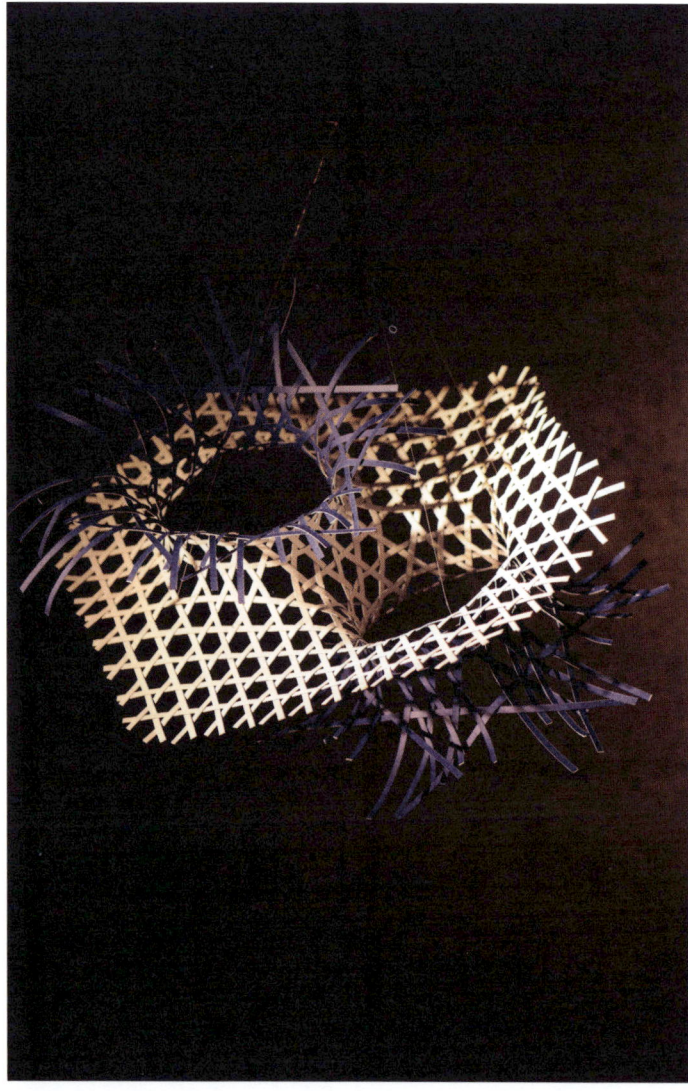

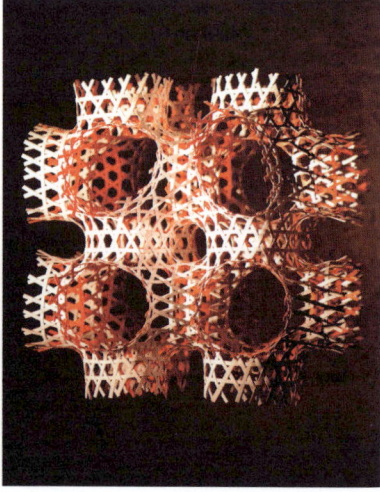

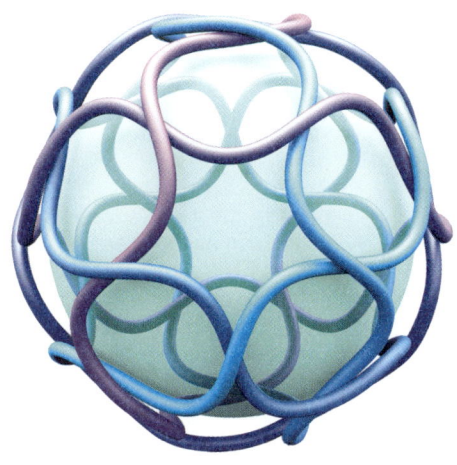 

↑　A digital rendering of a geometric design composed of filaments entangled over a sphere. The symmetry is close to that of the icosahedron, one of the platonic solids. Mathematically, the interest is to find rules for describing such structures.

↗↗　Rendering of a network whose topology is the dodecahedron. Its edges are tangled together, however, in a complicated way. The symmetry is close to the platonic dodecahedron, but contains only gyrations rather than mirror symmetries. This tangled structure challenges and extends the idea and definition of a polyhedron.

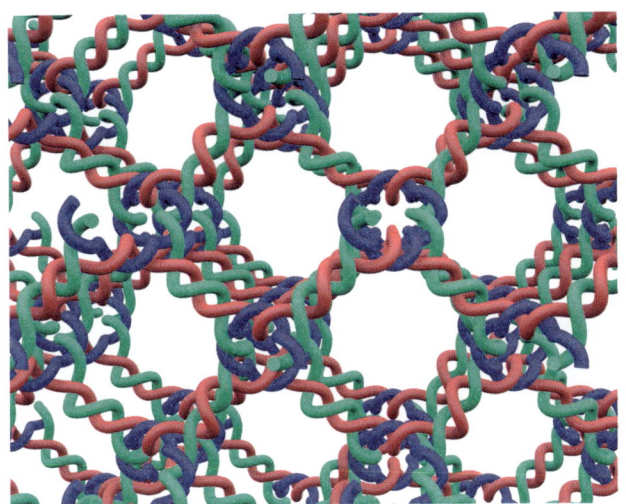

↗　An infinite array of tangling filaments in 3D, where the filaments weave in all three spatial directions rather than the two of a traditional fabric. Weaving like this could be used to make bulk material structures, particularly in nanotechnology.

→　An example of an entanglement of multiple honeycomb networks, which cover the whole plane to infinity. Each of the thirteen networks is related by symmetry to the rest.

we may see with the microscope. The other is the processes that have come about through evolution or evolved learning. If you take the idea of refining the technique of bundling material to make it stay together and abstract it into a geometric pattern, you're able to achieve more with less material.

ME:
The creative aspect for me lies in experimental mathematics. We know that there are specific surfaces and weavings that appear in biological systems. Mathematically, we can enumerate a far broader set of structures. The question is, do some of these hypothetical structures exist in other biological systems but we haven't had the tools to describe them yet? Mathematics allows me to step sideways and ask what other types of things I can do that are similar to what I've found. Also to explore further and ask what kind of functionalities I can get out of these other hypothetical structures. I think weaving and this intermediate space between topology and geometry are not well understood mathematically, and you just have to make them as 3D models on the computer to see what happens. What structures do I end up with? How are they related? Could I come up with a mathematical theory that covers all of them?

CS:
This bottom-up principle that you're describing is exactly how we work in our department of textile and surface design. We design by experimenting with materials and then end up with multiple prototypes on the table. We examine their properties—how strong or bendable they are, how they interact with light, etc. Based on this, we reflect on what they can be good for or what they're really good at, and then transfer them into an applied context.

AGM:
This explorative process of hypothetical structures is pretty much identical to my process of working on physical models. I start on something, and while making it, I realize that it would have been better to do it another way, or that an aspect I hadn't noticed would be more interesting. This process leads me to hypotheses of structures that I could possibly make. Generally, I'm in a state of frustration because while I'm working on something I'm already thinking about the other things that will be more interesting *after* that. But in a way, weaving works quite well. Once I've understood the structure, I'm just repeating the steps. It's like when you knit a complex pattern: at the beginning it seems really difficult, but after a few hours, you can do it and watch TV at the same time. The time I spend making a structure becomes good thinking time for understanding how to approach another problem. Having the physical object in my hand and being able to twist it around and see it from different

angles gives me a lot more information about a form.

EFW:
Repetition is a central concept to textile practice, and like you say, Alison, once you have the "script" defined for your weave it frees up time for reflection. Still, you always fill these scripts with material, and materiality plays such an important part that it will change the final structure. I'm curious to know if your work also has this material side, Myf, and if so, does it change the formulas?

ME:
Not really, but it is my goal to find connections between the entanglement of structures and their material properties. The tricky part is, which geometry would I choose for my entanglement? If I just have a simple knot, what configuration of that knot do I choose in order to explore its properties? How do I make this jump from something that's tangled together to a specific geometry that addresses functionality? Since I'm working in the very new field of nanomaterials, it's not clear yet what the functionalities can be. It's an exploratory process.

AGM:
For me, the constant issue is the materiality of the line. What is it made of? I imagine that for the ideal model, the line isn't going to have a dimension in space at all. If you make a physical model, you already have to compromise when that line becomes a rod. For me as a weaver, the line may have a section that's planar, because it's a piece of paper or fabric. If I'm trying to weave something more structural, however, I might use cylinders instead of rods, meaning I can connect them in some ways and not in others. So the materiality of the line really changes the possibilities of the model.

EFW:
In your computer simulations, Myf, your lines actually do have dimensions. How do you choose your specific modes of representation?

ME:
I've been working on it for a long time now and I've got different simulation techniques for various settings. My goal is always to find the optimized geometry of an entangled object, and what's optimal depends on what kind of setting you're working in. The first one I use makes the line into a tube with a fixed radius and then finds the geometry of the tube that has as little material as possible. I've worked quite a bit in periodic patterns as well, and I also have a project with a tubular structure modeled in a liquid. It's still very exploratory. It's like you were saying, Christiane: you need to just make them and then see how it goes.

EFW:
So these simulations are not only a form of representation, but a method of exploration and generating new knowledge?

CS:
And material thinking—embedding physical properties through theoretical thicknesses into your models. You're working with the idea of materialities, but mathematically.

ME:
Yes, absolutely. Materiality isn't something that a typical mathematician is interested in, because you've made a physical problem out of a geometric problem out of a mathematical problem. Alison talked about the process of putting things together, but it's something that's really difficult for my 3D structures. I've always struggled with how to put them together in a simple way. These new advancements in synthetic chemistry and molecular weavings are really exciting because they show that the mathematical processes I've been working on are reasonable in a real physical setting. The difference is that woven molecular structures aren't made by a weaver, but by putting molecular configurations that twist together into particular settings. That's a totally different process of weaving. This is a new thing for me and it's quite fun and really exciting.

EFW:
I think for most laypeople, weaving at a molecular scale remains very abstract and immaterial. Could you describe this process of molecules self-assembling into weaves and what it looks like?

ME:
We can't see the process itself, but we can design the pieces and we can draw a chemical diagram of the structure. Through crystallography or diffraction, we can work out what the nanostructure is, before and after. I find that this idea of self assembly is related to the process of textile making that Alison described earlier—that once you've defined the rules, you're just repeating the same thing over and over again. When the pattern is simple, you only have to train the molecules to do the same thing and they build the pattern. It's a nice connection to the idea of repeating patterns and programming that comes from weaving.

AGM
The simplicity of weaving is really interesting to me. By following rules of patterns and reiteration, complexity emerges from simplicity. In weaving, the elements stay in place through the friction of the over-under intersections, so there's no need for nuts or bolts or other binding material. And simple rules allow for easy constructions. If a structure is based on a modular approach and simple connections, then building can be done as teamwork—that's a great feature.

↑ The green bamboo is
split and the long strips
joined at crossings
to make a portable,
foldable structure.

↑↑ Woven bamboo sphere
in Alison Grace Martin's
garden in northern
Tuscany, where the
bamboo grows.

↑ Woven bamboo struc-
ture in preparation.

CS:
This idea of easy assembly through a repetitive pattern is really interesting. And as you say, everyone can learn it and it can become a shared practice. Could you both say something about what motivated you to seek out collaborations with other disciplines?

ME:
As a mathematician, I want to make my exploration as rigorous and broad as possible, which means that I need feedback from tangled systems in other contexts. I need to look at what chemists are doing in the lab. I'm also interacting with designers and architects, which is something new for me. It's quite interesting because designers and architects move much more quickly and freely. Science moves slowly: you can't proceed to the next step until you've completely understood the first. It's inspiring to see what people from other disciplines are doing, and then reflect on your own work: How does this fit into the framework of what I'm thinking about? Could these examples give me something to think about in that context? That's what all the different collaborations do for me. It doesn't really change what I'm doing, but it feeds into it. The more feedback you have into the system, the better it is.

AGM
For me, I suppose it has to do with knowing my own limitations and wanting to understand more. This came at a time when the internet was made accessible; it made a big difference. I did Google searches for hexagonal and pentagonal structures, and along with the woven things I was looking for, I found molecular shapes — it was around 2012, when graphene was a big thing. These images of simplified versions of molecular structures in scientific papers gave me an immediate understanding of how the shapes would work in three-dimensional space. So it wasn't coincidental, it was a time when the internet was something I had at home. The other thing that I've done a few times, and I'm afraid I did it with Myf as well, is to just cold-call people. People in academia leave their email addresses lying about; you can contact them and quite often they respond. Sometimes I've had really, really nice responses — the generosity of scientists has been unbelievable.

EFW:
In the community of weavers and textile crafts people, it's also always been very much about the idea of sharing knowledge and patterns with each other. With *Architectures of Weaving* we want to highlight the collaborative aspects of textile practice and it's a nice thought that the World Wide Web — essentially a virtual and material entanglement — enabled these personal "entanglements" so specifically.

Designing Performance

Material performance plays a central role both in natural and in man-made structures. Since textiles are considered an open material system — open to the integration of components of differing scales and materialities — they can be customized to specific needs. The fabric-bound fibrous elements can be of any scale from nano to macro and can range in materiality from processed mineral or metal to grown wood or wool. When combining such materials with differing textile construction methods like an elastic knit or a tight weave, a nearly infinite matrix of possible structural combinations opens up, each one with a distinct behavior resulting from the interplay of geometry and materiality. Thus the performance can be "programmed" into a textile through its enwrought constituents.

With nature as a role model for achieving specific functions through fibril-, fiber- or filament-based inner material structures, it is only a small step to exploring bioinspired design applications with textile practices. The functional programming can take place at the level of fibers, yarns or overall patterns. For instance, filament or yarn elements from hygroscopic cellulose (paper or wood) exhibit a shape change at microfiber level when humidified. When worked into a textile surface, this will cause a specific movement determined by the construction. "Bi-stable" systems are another common principle for shape change in nature that can be translated into textile technology when binding elements of differing inner strain together. Even fast-flipping movements from one form to another become possible with this design strategy.

In everyday life, engineered textiles are ubiquitous as "technical textiles," soft technical devices that are customized for highly specialized applications in fields like agriculture, landscaping, mobility, medicine, building, and many more. Recent examples are given in this chapter by a textile research institute. In the case studies you will find a resilient fabric designed to harvest water from clouds or a sun-shading system operating with interwoven Smart Materials. A traditional forerunner of today's technical textiles are the black tents of the Yörük nomads in Turkey which provide historical context, as well as contemporary points of reference.

Plant Fiber Twists

Team

Dr. Michaela Eder, Martin Niedermeier, Charlett Wenig, *MoA* Research Group *Adaptive Fibrous Material*, Max Planck Institute of Colloids and Interfaces, Department of Biomaterials, Potsdam-Golm, Cluster of Excellence *Matters of Activity. Image Space Material*, Humboldt-Universität zu Berlin

Context

Biological materials research, 2022

Material

Plant fibers — wood, hemp, cotton, imaged in the Low Vacuum mode of an Environmental Scanning Electron Microscope; drawings based on literature data related to cellulose orientation and fiber cross-sectional geometries.

The term "plant fiber" describes a single, elongated cell encased by a rigid cell wall. Fibers and fiber bundles play essential roles in the life of a plant, such as water supply, mechanical stability or as carriers during seed dispersal. The required specific functions of a type of plant fiber may change with time, with age or the size of a plant. In early spring, water transport needs to be efficient, which is realized by cells with large lumens (inner hollow space) and thin cell walls. Later in the growth season, more resources are allocated for the formation of cells with thicker walls and small lumens, resulting in denser, stiffer, and stronger fibers. In terms of mechanical stability, a young plant needs to be flexible to cope with wind loads; the trunk of a tall tree needs to be stiff to support its crown. Strong and dense fibers are also found in capsules or husks to protect seeds with their delicate embryos. On the other hand, lightweight and long fibers assist the wind-dispersal of seeds (*anemochory*) during their travel from the mother plant to a new location for germination.

The described multiple functions are achieved by the patterning of the cell walls: so-called cellulose synthase complexes, small cellulose-spinning "machines" located on the plasma membrane, deposit stiff and strong cellulose fibrils towards the outside of the cells. Vesicles, which are special transport compartments, transport lignin and hemicellulose building blocks from the cells' Golgi apparatus to the cell walls to incorporate them as a matrix between the cellulose fibrils. The alignment of the cellulose fibrils and the thickness of the cell wall strongly influence mechanical properties: fibrils oriented along the cell axis result in high-tensile stiffness. With the increasing angle of the cellulose fibrils to the longitudinal axis, the fibers become more flexible, which is favorable for young plants. So far, all the described

processes take place in a fully hydrated state. Upon drying, e.g. during fruit ripening or after harvest, the hygroscopic cell walls shrink mainly perpendicular to the cellulose fibril orientation. This results in cell twisting, again dependent on the orientation of the fibrils with respect to the cell axis as shown on p. 65. In the case of cotton fibers, these twists are not only crucial for seed dispersal via the wind but also highly beneficial for the usage of cotton fibers in textiles. It is well known that the distinctive fibril angles in plant cell walls — resulting from smaller subunits like microfibrils down to molecular arrangements — lead to fundamental fiber characteristics and properties.

Additionally, remarkable differences in cell lengths arise from cell synthesis and growth mechanisms — primary and secondary growth, elongation, tip growth — and can be easily observed. These differences are species-dependent and strongly influenced by growth conditions. While the effects of different cell lengths on the functionality in the organisms remain — at least partly — unclear, fiber lengths strongly determine the applicability of fibers for textiles and other products.

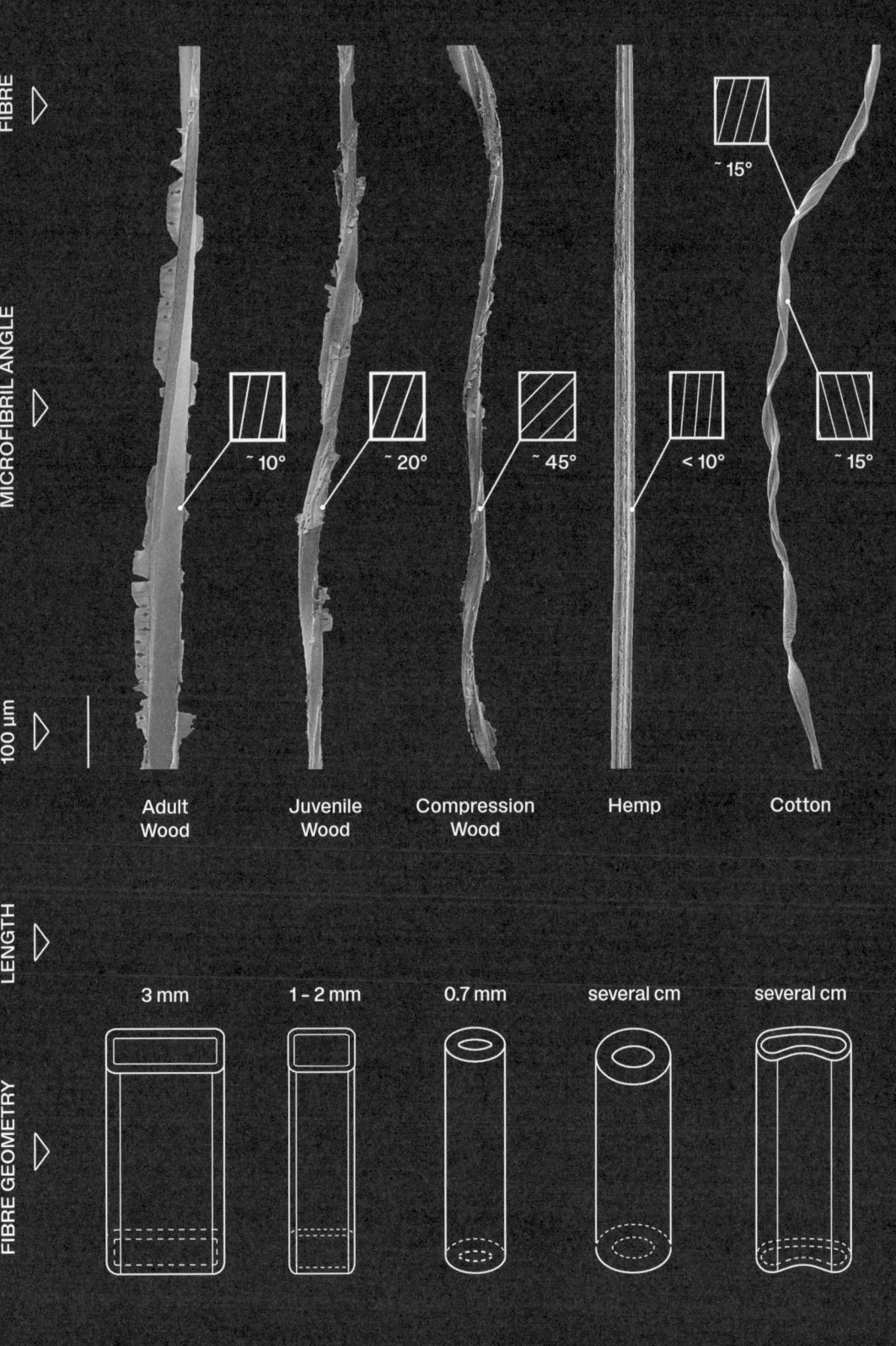

FIBRE

MICROFIBRIL ANGLE

~ 15°

~ 10°   ~ 20°   ~ 45°   < 10°   ~ 15°

100 μm

Adult   Juvenile   Compression   Hemp   Cotton
Wood   Wood   Wood

LENGTH

3 mm   1 – 2 mm   0.7 mm   several cm   several cm

FIBRE GEOMETRY

Biomaterials and Design
Peter Fratzl, Christiane Sauer

WHY (BIO)MATERIALS SCIENTISTS, ARCHITECTS, AND
TEXTILE DESIGNERS NEED TO INTERACT MORE

It is obvious that architects need to know the properties of materials for any construction planning. In this sense, they build upon the work of materials scientists. Any new material class with better properties will have an impact on the architectural design process, for example by allowing more freedom for novel structures by increasing stability and by providing a new idea of construction. For instance, the development of steel as a standardized building material has fundamentally changed building typologies because its tensile strength enabled tall constructions and large spans. Steel constructions are hierarchically structured systems composed from isotropic linear elements with consistent properties. But what if even the material could be designed to become a structural system in itself? Could we think of building components that adapt to outer conditions and seasonal changes through growth or change like those we find in nature's structures?

In recent years, materials scientists have started to talk increasingly about architecture in the context of material structure.[1] The reason why is that material properties are largely dominated by their micro- and meso-scopic structure. With the recent development of micro- and nanotechnological fabrication techniques, it has become more and more practical to control the internal architecture of materials and, thereby, the physical properties without changing their chemical composition.

This is exemplified in Fig. 1, which refers to the conception and construction through architecture of wooden buildings reaching heights close to 100 meters; but wood itself is a complex architectural arrangement of micrometer-sized tubes and nanometric fibers wound around them. What we call wood material is just what happens to be visible in the centimeter range that is most accessible to us humans. In fact, a wood log is a complex architectural construction in its own right that emerged through evolution and serves the living plant as structural support and for water and nutrient transport.

1 Y. Estrin et. al., "Architecturing materials at mesoscale: some current trends,"
Materials Research Letters 9 (2021): 399–421.

It is therefore logical that materials scientists have to rely increasingly on mechanical principles that are also the basis of architectural construction to even understand a natural material such as wood. Furthermore, various kinds of architectured materials (materials whose properties are determined by their inner structure) will start to emerge as new members of all materials classes, metals, polymers, and ceramics. The concept of architectured materials is not new and the most traditional representatives of that class are textile structures that can be considered three-dimensionally engineered fiber structures. Such materials provide various and unusual property combinations[2] acquired through their architecture, and can be lightweight and adapt the number of material combinations required to reach certain functionalities. Grading is readily introduced into textile construction and into architectured materials in general. In this respect, architectured materials are akin to composites based on joining components with different properties. If one of the system's components is just void, the result effectively is a mono-material system, which allows for easy recycling processes. In addition, a textile system joins its material components, threads, and fibers, merely through loose resolvable entanglements and can therefore be disassembled after its cycle of use. Thus, unlike glued composites, architectured materials have the potential to impact the sustainability and recyclability of materials in a positive way.

Fig. 1
Multiscale architectures.
Reprinted from P. Fratzl, C. Sauer, and K. Razghandi, "Editorial for a Special Issue on Bioinspired Architectural and Architected Materials," *Bioinspiration and Biomimetics* (2022), https://doi.org/10.1088/1748-3190/ac6646. Left: The wooden tower of lake Mjøsa, 2022 (with permission from Voll Arkitekter AS + Ricardo Foto) Right: the internal structure of wood based on parallel tube-like wood cells with diameters in the range of tens of microns (shown for several wood species from top to bottom). The white arrow points to a sketch where nanometer-thick cellulose fibrils are indicated by black lines. Adapted from M. Eder et al., "Wood and the Activity of Dead Tissue," *Advanced Materials* 33 (2020): 2001412.

Architecture based on wood

Wood based on (internal) architecture

Architects and designers may profit from knowledge gained from the observation and understanding of (natural and artificial) architectures in the microscopic world and transfer them to the scale of buildings. But in architectural practice today, design, materiality, and construction are often treated as separate subjects, despite significant efforts to overcome this.[3] First, a design is created, then a desired material is chosen, and then an engineer is consulted to specify dimensions and geometries. In nature, it is the other way around: material, structure, and form are an inseparable entity, as seen in the example of wood. The material logic generates form to meet a specific purpose. This approach would be considered a "bottom-up" principle in design: materiality shaping the design—a

2   M. Ashby, *Materials Selection in Mechanical Design* (5th Edition) (Cambridge, MA: Elsevier, 2016).

3   M. Bechthold and J. C. Weaver, "Materials science and architecture," *Nature Reviews Materials* 2 (2017).

practice we can still find in skilled craftmanship that is aware of the material's inner activity and therefore works *with* it. This puts material at center stage. As architects, we should again dive deep into the architecture of materials to be inspired by nature's potential for intelligent design solutions.

Charles and Ray Eames' impressive film *Powers of Ten*[4] visualized in 1977 the merging of scales from atom to body, to city, to earth, and to universe. The short video elucidates that none of the levels can be approached separately, since each scale is always simultaneously part of a larger and of a smaller system. This showcases the interdependency of material systems and gains new actuality today. Built architecture can avoid neither the material's inner activities nor the environment's outer influences. For the creation of resilient approaches that will meet the more extreme and demanding conditions of the coming decades, we rely on understanding nature's contexts in order to translate them into sustainable improvements for our built and non-built environment.

NATURE PRINCIPLES

Natural materials with a wide range of properties have emerged through evolution. Their functional diversity is primarily a result of their hierarchical structure,[5] because they are essentially based on a similar set of base materials: proteins, polysaccharides (mostly

Fig. 2
Examples of biological materials based on proteins, sugars and/or minerals. Reproduced with permission from M. Eder, S. Amini, and P. Fratzl, "Biological composites-complex structures for functional diversity," *Science* 362 (2018): 543–47.

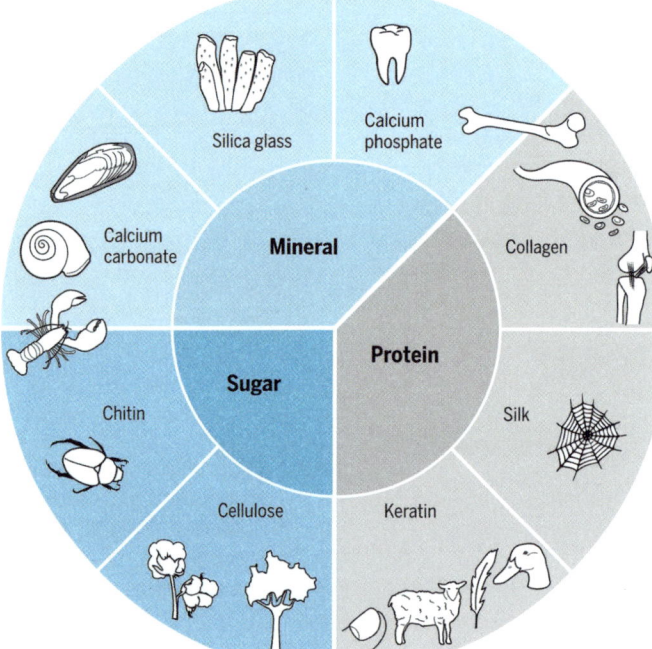

cellulose and chitin), as well as some minerals. This includes diverse objects, such as silk fibers in spider webs, byssus fibers that attach mussels to rock, collagen fibers in skin, tendon or bone, cellulose fibers in plants or bacterial biofilms or chitin fibers in insect cuticles.

4   C. and R. Eames, 1977, "Powers of Ten" © 1977 Eames Office LLC. Accessed April 1, 2022, https://www.youtube.com/watch?v=0fKBhvDjuy0.
5   P. Fratzl and R. Weinkamer. "Nature's Hierarchical Materials," *Progress in Material Science* 52 (2007), 1263–334.

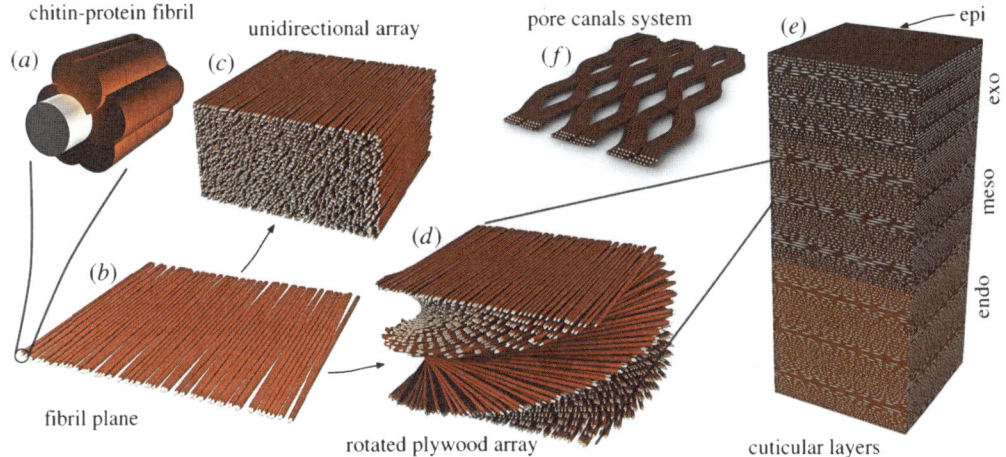

chitin-protein fibril
unidirectional array
pore canals system
epi
exo
meso
endo
(a)
(c)
(f)
(e)
(b)
(d)
fibril plane
rotated plywood array
cuticular layers

Fig. 3
Chitin fiber architectures in the spider cuticle. Reproduced from Y. Politi, L. Bertinetti, P. Fratzl, and F.G. Barth, "The Spider Cuticle: A remarkable Material Toolbox for Functional Diversity," *Philosophical Transactions of the Royal Society A: Mathematical, Physical and Engineering Sciences* 379 (2021): 20200332.

Many examples of materials based on (extracellular) protein fibers are given in a recent review.[6]

As an example of an architecture of natural materials based on fiber arrangements, Fig. 3 shows structures found in different parts of the spider cuticle that confer very different properties, such as compression strength, tension stiffness, color, and even transparency.

"Building" with fiber architectures leads to a huge variety in nature's toolbox. If we want to utilize such structural principles in design, the idea cannot be to exactly imitate nature because we have to transfer materiality into another scale and context, but to be inspired by its concepts. Fig. 4 mentions some principles of natural material structuring and synthesis as compared to traditional engineering principles. Topics such as structural hierarchies, adaptation, and self-repair are currently investigated as new paradigms in materials design.

This approach differs from classical engineering, where material deployment is calculated for a theoretical maximum impact and multiple security factors leading to over-dimensioned systems in everyday use. If we aim to transfer biological systems as an inspiration into design, then material hierarchy, adaptivity, and flexibility will play a central role. These properties can be found in textiles that allow to program performance by architecting few components into a complex fiber structure — like those found in biological material. Therefore, textile-based design will gain growing importance in developing sustainable solutions with nature as a role model.

FIBER ARCHITECTURES

In textile fabrication we can program performance on all levels of fiber, yarn, textile structure, and shaping. As in nature, this hierarchy in composition can result in an immense variety of performance, since the structural hierarchy can be varied and designed to specific needs.

For instance, the same woolen fiber can be processed for different outcomes: as a loose staple fiber it can be felted into a stiff nonwoven sheet, as a spun yarn it can be woven to become a strong tensile cloth or it can be knitted to form an open elastic surface. In addition, the natural micro- and nanostructure of wool induces

6   M. J. Harrington and P. Fratzl, "Natural Load-Bearing Protein Materials," *Progress in Materials Science* 120 (2021): 100767/1–44.

Fig. 4
Differences between bio-
logical and traditional
engineering materials.
Reproduced from Y. Estrin,
Y. Beygelzimer, R. Kulagin,
P. Gumbsch, P. Fratzl, Y. Zhu,
and H. Hahn, "Architecturing
Materials at Mesoscale: Some
Current Trends," *Materials
Research Letters* 9 (2021):
399–421. Adapted from
P. Fratzl, "Biomimetic Material
Research: What Can We
Really Learn from Nature's
Structural Materials?", *Journal
of the Royal Society Interface*
4 (2007): 637–42.

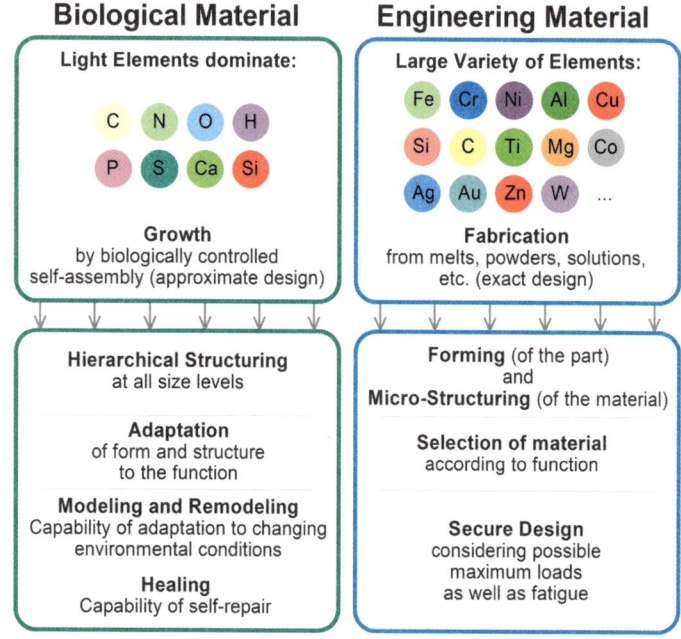

hygroscopic behavior and therefore causes the resulting fabric to absorb humidity. This property is applied, for example, in the woolen weaves of traditional nomad black tents. They are woven from local goat hair and utilize the hygroscopy of the fiber to shield the interior from rain. The wetted woolen yarn of the open weave swells and creates a closed saturated surface, which generates a coherent water film guiding further raindrops directly to the ground.[7] In this case, rain itself becomes part of a barrier to protect the space from water. The natural fiber properties in combination with the structure of the weave create an adaptive material system that becomes a protective membrane responsive to the environment.

Fig. 5
The hierarchy of textiles.
Reproduced with permission
by Wiley from V. Sanchez,
C. J. Walsh, C. J. and R. J.
Wood, "Textile Technology for
Soft Robotic and Autonomous
Garments," *Advanced
Functional Materials* 31
(2021): 2008278, https://doi.
org/10.1002/adfm.202008278.

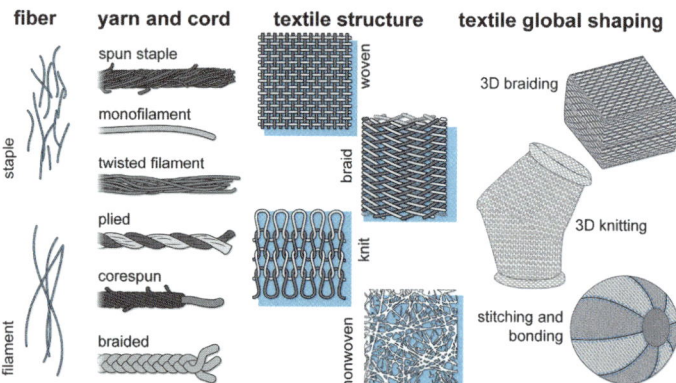

Material grading is another functional parameter found not only in natural but also in man-made material systems. A textile can vary its pattern and therefore grade its density or shape. Knits can come off the machine fully fashioned in a finished three-dimensional shape without any cut-offs—a material-saving and structurally effective means of production. A well-known example is warp-knitted tights with a pattern-induced strengthening at heels, toes, and hips. A weave, on the other hand, can be graded by combining differing materials in warp or weft. Threads of different elasticities, for instance, result in a three-dimensional buckling and shaping of the surface (Fig. 6). When integrating rigid elements like sticks or rods, textiles can even gain stability, or when embedding conductive fiber, they can act as information-transmitting devices, as sensors or actuators (e-textiles). This shows that textile structures are highly suitable for designing functional material systems with programmable properties and performance. The fields of applications seem endless, and range from lightweight architectural structures, customized upholstery and furniture, climatic and acoustic interior elements to soft robotic devices for motion-assistive garments.[8]

MATERIAL ACTIVITY[9]

The Research Cluster *Matters of Activity. Image Space Material* started in the year 2019 with the intention to bridge disciplines for developing a new, sustainable culture embracing the activity of material.[10] It facilitates the collaboration of materials scientists with architects and designers who would never have met otherwise, since they work in different places such as university, research institution or art school. Through transdisciplinary research and prototyping, novel disciplinary perspectives for research and applications are opening up.

Fig. 6
Left: Graded Yarns, Elisabeth Bauer from "InVisible the Sting" (2021): Master Thesis, Weißensee School of Art and Design Berlin, mentors C. Sauer, J. Petruschat.
Right: Multi-layered graded Jacquard weave binding differing fibers, Samira Akhavan from *Dis/cover* (2019): MoA Design Studio *Scaling Nature (1) Wrinkles*, Weißensee School of Art and Design Berlin, mentors C. Sauer, E. Fransén Waldhör, M. Schneider, K. Razghandi, L. Guiducci.

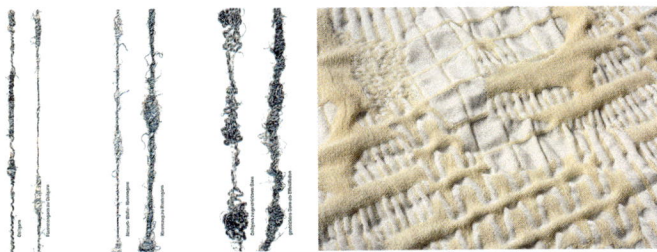

As an outlook of this short plea for a more systematic and programmatic collaboration between architects and (bio-)materials scientists, we show one example of a design study rooted in this Research Cluster, making use of the hygroscopicity of wood and its internal structure, which codes for specific complex macroscopic structures and movements. Indeed, natural material allows to utilize fiber's inner activity, especially for environmental interaction. Cellulose-based material responds to variations in humidity by changing shape—obvious in the wrinkling of humid paper. Wood as a ubiquitous cellulosic material in design and building is deployed in common practice despite its inner hygroscopic activity, not because of it. Deployed constructions or coatings intend to keep water away and maintain machine-cut geometries while neglecting the material's inner structure. The wooden cell walls are composed of cellulose fibrils that swell with humidity and trigger a specific movement dependent on the fibril orientation.[11] If wood is allowed to unfurl this inner activity, it becomes an environmentally responsive building material.

8   V. Sanchez, C. J. Walsh, C. J. and R. J. Wood, "Textile Technology for Soft Robotic and Autonomous Garments," *Advanced Functional Materials* 31 (2021): 2008278, https://doi.org/10.1002/adfm.202008278.

9   *Active Materials*, ed. P. Fratzl, M. Friedman, K. Krauthausen, and W. Schäffner (Berlin: De Gruyter, 2022).

10   The Cluster of Excellence *Matters of Activity. Image Space Material* at Humboldt-Universität zu Berlin brings together more than forty disciplines to systematically investigate design strategies for active materials and structures that adapt to specific requirements and environments. See also, the Preface in this publication, p. 11.

11   I. Burgert and P. Fratzl, "Plants control the properties and actuation of their organs through the orientation of cellulose fibrils in their cell walls," *Integrative and Comparative Biology* 49, no. 1 (2009): 69–79, https://doi.org/ 10.1093/icb/icp026. Epub 2009 May 22.

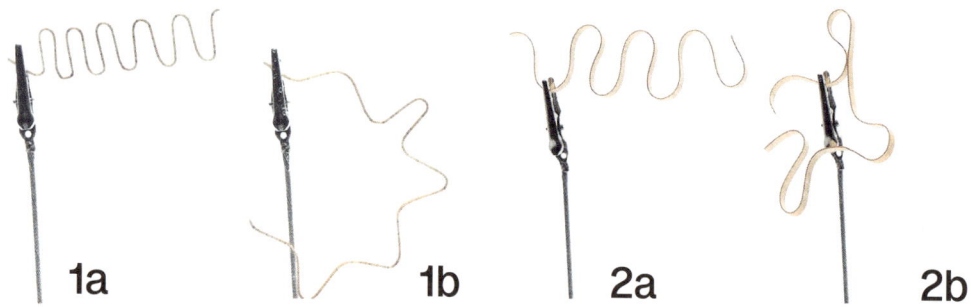

1a 1b 2a 2b

Fig. 7
Hygroscopic shape change of preformed maple veneer strips, cut with the grain (1a, b) and cut against the grain (2a, b). From *Living Beings*, Nelli Singer, Weißensee School of Art and Design in cooperation with *Matters of Activity. Image Space Material.*

The design research project *Living Beings* by Nelli Singer (see p. 90) evolved from the transdisciplinary exchange between textile design, architecture, and biomaterial science.[12] This design research showcases thin wooden veneer strips as hygroscopic, active "yarns" that change shape depending on wood type, strip dimension and angle of fiber direction. Based on these parameters, numerous patterns were analyzed for their performance through experimental prototyping. The knit structures exhibit reversible shape change that can be programmed by varying the design parameters on every hierarchical level of the textile construction. Linear strips of veneer are preformed into loops before being interlocked by hand into a flexible surface. When wetted, the strips exhibit shape-memory behavior and stretch out towards their initial shape (Fig. 7). The multiple fiber motions add up and cause the overall structure to bend, expand or contract with changing humidity levels (Fig. 8). Upon drying, the movement is reversed. The resulting lightweight screens can alter their density and shape. They could become adaptive devices for climatic modulation in a building skin or an interior setting—activated directly by the environmental conditions. Beyond being a self-sustaining technical mechanism that operates without any electrical input, the project fascinates because of its delicacy of design and an animated motion that evokes curiosity and wonder—an appealing combination that is needed to transfer such novel approaches into products and societal awareness (Fig. 9).

12 N. Singer, *Living Beings* (2020): Master Thesis, Weißensee School of Art and Design Berlin, in cooperation with *Matters of Activity. Image Space Material* mentors C. Sauer, J. Petruschat, M. Eder, L. Guiducci, and K. Rhazghandi.

Fig. 8
Hygroscopic knitted structure,
maple veneer, 1 mm, 3 mm,
stitch size 6 mm. Shape-
change duration 3:36 mins.
From *Living Beings*, Nelli
Singer, Weißensee School of
Art and Design in cooperation
with *Matters of Activity. Image
Space Material.*

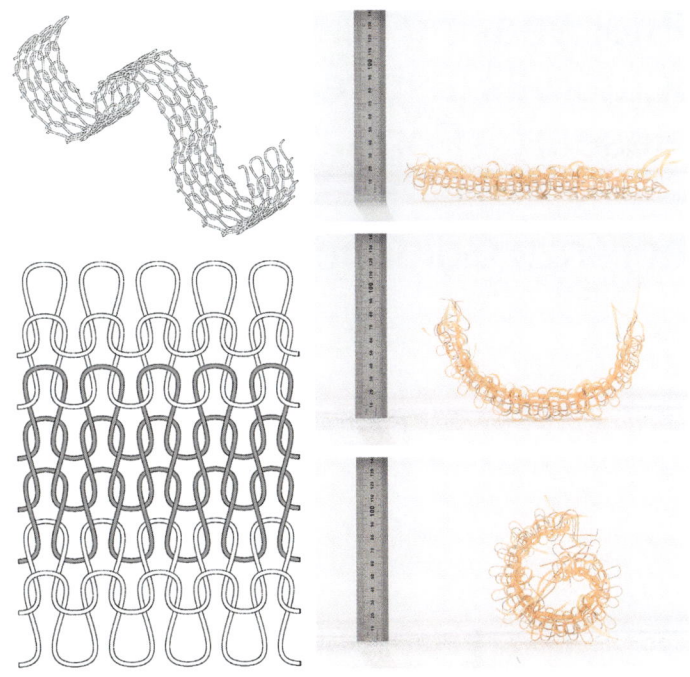

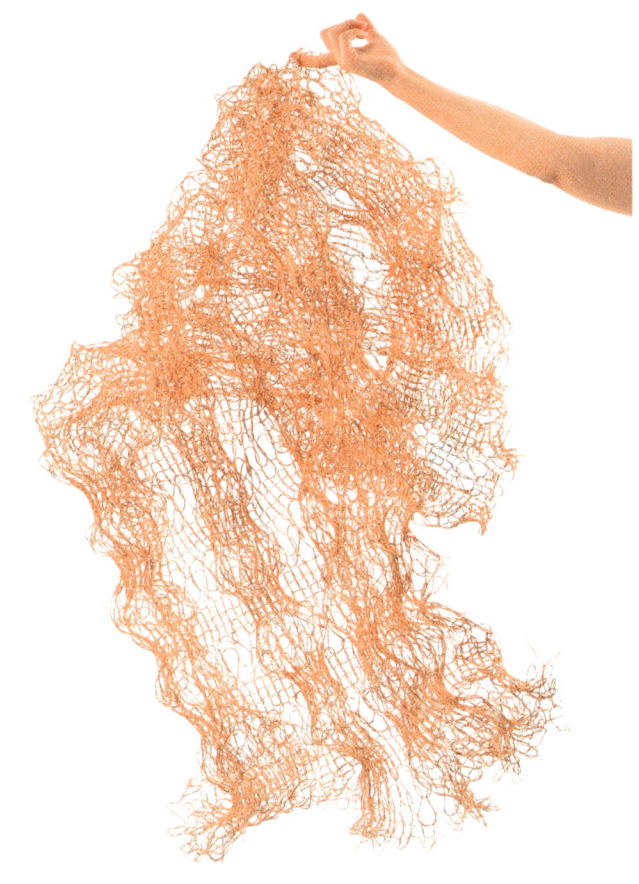

Fig. 9
Hygroscopic knitted surface
125 × 182 cm. Maple veneer,
1 mm, stitch size 10 mm,
20 mm. From *Living Beings*,
Nelli Singer, Weißensee
School of Art and Design in
cooperation with *Matters of
Activity. Image Space Material.*

Interweaving Disciplines—How Biological Materials Inspire New Teaching and Research Formats in Design

Lorenzo Guiducci

INTERDISCIPLINARITY IN NATURAL SCIENCES: THE EXEMPLARY CASE OF BIOMIMETICS

Today's academic world has been taken by storm by the advent of interdisciplinarity. Answering disciplinary questions using the tools of another discipline is a success story. In natural sciences the nexus of different techniques, experimental practices, objects of study, have transcended the traditional disciplinary boundaries, creating new epistemologies and communities of peers.

One such success story is *biomimetics*, the process by which knowledge about a given biological system is transferred to the technical realm to solve a difficult problem in a new, alternative and "smart" way. In fact, biomimetics can be said to follow two main directions: the "biological push," by which new insights from a biological system suggest a new technical process/application/product, and the "technological pull," by which already studied biological systems are mined to find a solution for an existing technical problem. According to this "utilitarian" interpretation, biomimetics satisfies two very different goals: on the one hand natural scientists study a biological role model to extract new *knowledge*; on the other hand, engineers are more focused on finding a viable *solution*.

THE DISTANT WORLDS OF DESIGN AND NATURAL SCIENCES

Design and architecture focus on the physical interaction between humans and the technical, built, artificial world. As in engineering, design and architecture also aim to find solutions. Hence, it doesn't come as a surprise that even in a design school, courses and teaching formats favor exchange with engineers, industrial and technology partners. In such collaborations, designers aim to envision new products and technologies, identifying emerging needs and social demands. In contrast, collaborations between designers and natural scientists are not common, and the two worlds, especially at the academic level, remain quite distinct: while scientists work with complex and exact tools (both theoretical and experimental), designers traditionally master crafts and hands-on techniques; while physical making is a primary practice and founding aspect of design, observation and verification is the typical *modus operandi* of scientists.

Given these premises, how do we reconcile the practical nature of design with a scientific foundation that goes beyond industrial-age categories? Which interdisciplinary tools do today's design students need that will be relevant in the future? Can new research questions be rooted in both design and fundamental science? Which relevant outcomes can be envisioned from an integrated interdisciplinary approach, besides a physiological increase in productivity deriving from sharing methods? And if so, for whom — for the designer, for the scientist or for both?

A TEACHING FORMAT WEAVING TOGETHER BIOLOGICAL MATERIALS SCIENCE AND TEXTILE DESIGN

These are daunting questions that cannot be treated in a complete manner in this short essay; above all, they prompt non-univocal answers, since various new forms of collaboration between design and scientific disciplines are being experimented with in both research and teaching. Nevertheless, we would like to provide a possible perspective on the topic, not by means of generalizations, but by reporting our personal experience (as "fieldwork," so to speak) of a new teaching format we conceived, resulting from interdisciplinary collaborations taking place at the Cluster of Excellence *Matters of Activity. Image Space Material* as "fieldwork" *(MoA).*

The proposed MoA Design Research Studio *Scaling Nature*[1] was developed over three semesters from 2019–20 and targeted at Bachelor and Master students of Textile and Surface Design at the Weißensee School of Art and Design Berlin. Since the department is especially focused on material design with fiber-based and textile techniques the research bears a strong relation to the structure of biological tissue. Specific natural phenomena introduced by scientists were taken as a source of inspiration for exploring novel approaches in the field of sustainable and adaptive design concepts. The structure-function relationships ranged from micro-scale in biomaterials to macro-scale in *Gestaltung* through physical prototyping in an experimental way. In turn, the findings gained through the hands-on modeling of the designers ignited new research questions and perspectives for the scientists. A fruitful process of exchange, learning, and development started, through stitching together the disciplines.

WHY DID WE DO IT?

Our primary motivation was to conceive *Scaling Nature* as a means to gain a more thorough understanding of nature and especially biological materials by design students who usually do not get to study materials from this point of view. With *Scaling Nature*, we wanted the students to go beyond the aesthetic appearance of nature and get inspired by "nature's inner workings," taking part in the same excitement that scientists experience when investigating their objects of study.

In this regard, a related consideration is the *praxis* of biomimetics: as we briefly mentioned earlier, the actors involved in extracting/transferring knowledge from a biological system to a technical domain are typically natural scientists on one side and engineers on the other; such a target-oriented "from expert to expert" strategy implies a strict adherence to a disciplinary subdivision of

1   Karola Dierichs, Ebba Fransén Waldhör, Lorenzo Guiducci, Agata Kycia, Khashayar Razghandi, Iva Rešetar, Christiane Sauer, Maxie Schneider, Mareike Stoll.

Fig. 1
Top: Assembly of two *Sticky Stacks* building blocks.
Bottom: A hanging ribbon built with many such blocks, showcasing resistance to tensile loads.

competences. On the other hand, designers typically work by transgressing borders when creating new solutions.

With *Scaling Nature*, we therefore wanted to favor mutual exchange between scientists and designers by swapping their roles as experts in their own discipline and laymen in the other, opening up a space for (serendipitous) discovery leading to unforeseen results. Exposed to the analytical work of biomaterials experts, students gained a thorough understanding of the biological systems and their mechanisms, obtaining an increasingly profound knowledge throughout the Design Studio. In parallel, scientists discovered novel aspects of their research in the physical prototypes based on the free — even artistic — interpretations of the biological role models.

HOW DID WE WORK TOGETHER?

Each of the three *Scaling Nature* Design Research Studios focused on a different topic related to biological materials: *Wrinkles* looked at the biological formation of thin sheets such as bacterial biofilms, skin, corals; *Fibers, Muscles, and Bones* dealt with the phenomena of force generation and structural support in organisms; *Growth*, as the biological process *par excellence*, was presented not only as a mechanism for accretion but also for specialization and programmability of biological tissues.

These topics were chosen from the ongoing scientific research at the related institutions[2] for their relevance and transferability to overarching concepts such as activity, adaptability, responsiveness. At the beginning of each Design Studio, these scientists, invited as guest lecturers, provided a theoretical primer about their object of study and the techniques used. Through the lenses of biology, physics, mechanics, and materials science, a strong emphasis was put on highlighting how a material's structure informs its multiple functions in an organism.

The insights into material structure, growth process, biological control or form-induced functionality became a source of inspiration for the students to be developed further in iterative design circles. An important aspect was the analytical approach towards physical prototyping throughout the Design Studio. The students would explore possible design solutions and fabricate new proto-

2   Namely, Cécile Bidan, Mason Dean, Michaela Eder, Peter Fratzl at the Biomaterials Department of the Max Planck Institute of Colloids and Interfaces, Potsdam/Golm, and John Nyakatura at the Faculty of Zoology at Humboldt-Universität zu Berlin.

types in a systematic way: most often, geometric, material and process parameters were varied in a certain range. Only after observing the resulting prototypes and their performances (broadly speaking) would the students focus on a smaller, specific set of these parameters, or change the baseline design completely, adopting a different one. Group discussions, accompanying workshops, and supervision from lecturers supported the design process, which mainly applied textile-related manufacturing (digital weaving, automated knitting, 3D printing on textile, etc.). Through presentations and constant exchange, each design project underwent successive stages of broadening and focusing, not fixated on a predetermined outcome.

A VARIETY OF DIFFERENT OUTCOMES

This freedom was heavily reflected in the large variety of concepts that became tangible design prototypes: furniture pieces, material developments, textile or yarn-making processes, adaptive devices, and modular construction systems.

Besides the actual physical outcomes, the students' research work resulted in fortuitous discoveries too: this is the case, for example, with the *Sticky Stacks*[3] project, in which the students took the parallel arrangement of actin and myosin molecules in the skeletal muscles as role models for force generation. An upscaled, macroscopic mock-up of the sarcomere was built, using stacks of paper sheets interleaved together. Surprisingly, two paper stacks assembled in this way were found to be extremely difficult to pull apart. Just as tiny forces generated biochemically are multiplied by the sheer number of parallel actin-myosin units, the modest friction between two sheets was exponentially multiplied thanks to the interleaved arrangement. Structures built with such paper building blocks obtained extremely high strength, although only in terms of tension (Fig. 1).

As a result, the students project called for a scientific deepening that has led to ongoing research at the Cluster of Excellence *Matters of Activity*, focusing on methods to tweak friction in interleaved sheet assemblies, with the goal of obtaining sustainable paper-based structures with a load-bearing function (see *Woven Paper Bridge* Case Study in this book, p. 41).

The project *Hydroweave*[4] took inspiration from hygroresponsive plant structures (e.g. pine cones) that react autonomously to environmental humidity. The orientation of cellulose fibrils in plant cells generate directional stresses, thus enabling programmable deformation patterns upon changes in humidity. In a similar way, artificial paper yarn unfolds in different directions upon swelling, depending on its twist. Leveraging this property, sheets loosely woven from paper yarn were designed as room dividers with autonomous shape-changing triggered by environmental humidity (see also *Hydroweave* Case Study p. 81).

In *The Flip*,[5] the spore dispersal strategy of mushrooms was the main inspiration driver. In this case, the mushroom cap greatly shrinks upon drying, thereby exposing the lamellae and dispersing the spores. To explore this phenomenon, an analog design model was developed using pre-stretched fabric, onto which hot glue was deposited in a pattern of radiating ridges, inspired by the mushroom gills. Upon tension release, the fabric would take on a roughly toroidal shape. Interestingly, thanks to the residual elastic energy in the fabric, a little pushing in the center of the radial

---

3   Design by Serafina Baucken, Eva Eckert, Josephine Shone.
4   Design by Juni Sun Neyenhuys, Stefanie Eichler.
5   Design by Xingwen Pan.

78

Fig. 2
The Flip, a mushroom-inspired
hybrid textile snap-closing in a
spherical capsule

pattern would flip the fabric into a roughly spherical capsule. Here, a physical phenomenon completely different from the starting biological system—snap-through instabilities upon curvature inversion—was "discovered" (Fig. 2).

Eventually this project was deepened as a Bachelor Thesis by the textile designer (see *Skeletal Surface* Case Study p. 84) leading to an artistic production that expresses the sculptural plasticity of textile hybrids. In parallel, the potential of combining a rigid material with a soft pre-stressed one was further explored in a collaboration between two of the lecturers—notably, an architect and a biomechanical engineer—focusing on pre-stretched fabrics covered with 3D-printed rods of rigid plastics. The resulting textile hybrids served as boundary research objects for both architectural and biomechanical studies that have been published in corresponding disciplinary publication outlets.[6] While the architectural study demonstrated the potential uses of such textile surface elements for building skins, the biomechanics study provided an analytical model relating the different obtainable morphing patterns to the geometric, mechanical, and material parameters of the textile hybrids (Fig. 4).

6   A. Kycia and L. Guiducci, "Self-shaping Textiles—A material platform for digitally designed, material-informed surface elements," paper presented at eCAADe Conference, TU Berlin, September 2020; L. Guiducci, A. Kycia, C. Sauer, and P. Fratzl, "Self-organized rod undulations on pre-stretched textiles," *Bioinspiration & Biomimetics* 17, no. 3 (May 2022), https://doi.org// 10.1088/1748-3190/ac5b85.

After three semesters of teaching the *Scaling Nature* Design Research Studios, we drew some conclusions, which we briefly report here.

The hands-on approach and analog models developed during the studio proved successful in bridging disciplines through "boundary objects" useful to illustrate complex theoretical concepts (like elastic instability, strain energy, etc.), in a similar way to what was experimented with in the Bauhaus school.

Moreover, the explorative work of design solutions was tracked both in visual and written form (Fig. 3), producing a record not dissimilar to lab-books by experimental scientists. This common approach highlights the strong contact points between science and design and their intrinsic practices.

For the scientists involved in the lectures, the experience meant more than simple scientific communication and outreach. The "free" interpretation of geometric construction principles borrowed from exemplary biomaterials provided cues to new research ideas, challenged them to engage with new experimental setups at larger scales, and increased the relevance of their research outside the disciplinary boundaries.

For the design students, this experience enriched their empirical intuition of material behavior, leveraging knowledge already gained through past experience to translate a biological principle to the macro-scale.

In short, the overall experience has proved how the supposed gap between fundamental research in natural sciences and applied research in design disciplines can be naturally filled—provided that room and time is made to allow for such collaborations to unravel. Indeed, not only has the hybrid team established a common ground leading to science-design research collaborations beyond teaching, but the work of students developed during each semester has nucleated ideas to be expanded upon and researched in the coming years. Design and science entered a novel space for transdisciplinary collaboration. In contrast to a deductive target-oriented

Fig. 3
Material studies for *The Flip*, showcasing the systematic exploration of ridges geometry and adhesive types.

—— glue on front side
—— glue on back side

—— glue on front side
—— glue on back side

—— glue on front side
------- glue on back side

d=30 cm

—— glue on front side
⟹ glue line: from thin to thick

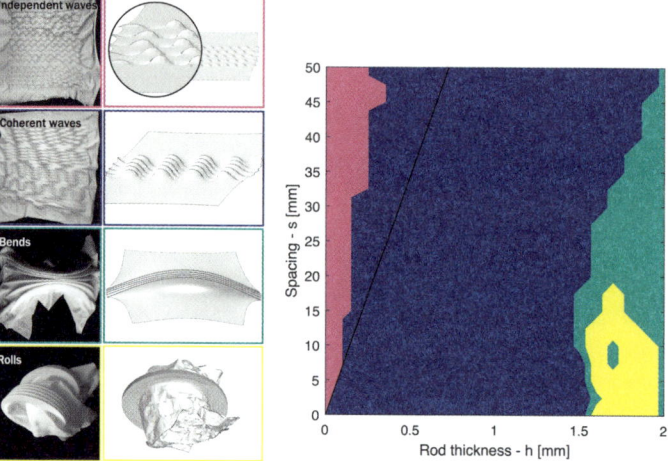

Fig. 4
Scaling Nature worked as an incubator for science-design interdisciplinary collaborations. Left: Physical and finite element models of a hybrid textile composed from straight rods, 3D-printed on a pretensioned membrane; Right: Phase diagram of the morphologies shown on the left.

design approach, our mutual design-inspiration in science and of bioinspiration in design leads to a continuous thread of collaboration and research. The borders between experts and laymen, between teachers and students are dissolving with the practice of learning from each other.

This approach to teaching, strongly supported by free experimentation, allows for disruptive innovation since no intended outcome is formulated a priori. A continuous broadening and focusing becomes possible, allowing one to stumble upon interesting and surprising observations, resulting in advanced design and innovative applications that include multiple perspectives and reflect nature in a multifaceted way.

Hydroweave

Team

Stefanie Eichler and Juni Sun Neyenhuys, Weißensee School of Art and Design Berlin, Department of Textile and Surface Design
In collaboration with Cluster of Excellence *Matters of Activity. Image Space Material*, Humboldt-Universität zu Berlin and Max Planck Institute of Colloids and Interfaces Potsdam/Golm, Department of Biomaterials
Mentoring team: Prof. Christiane Sauer, Ebba Fransén Waldhör, Maxie Schneider, Dr. Lorenzo Guiducci, Dr. Khashayar Razghandi

Context

Design Prototype, MoA Design Research Studio, *Scaling Nature (1): Wrinkles*, 2020

Material

Cellulosic paper yarn, twill woven modules, support netting

↓ Textile design parameters for programming movement.

diameter of the
yarn: 0.6 mm

direction of the
twist: Z

warp threads
per cm: 3

binding: twill

size: 19 × 19 cm

subconstruction:
mesh

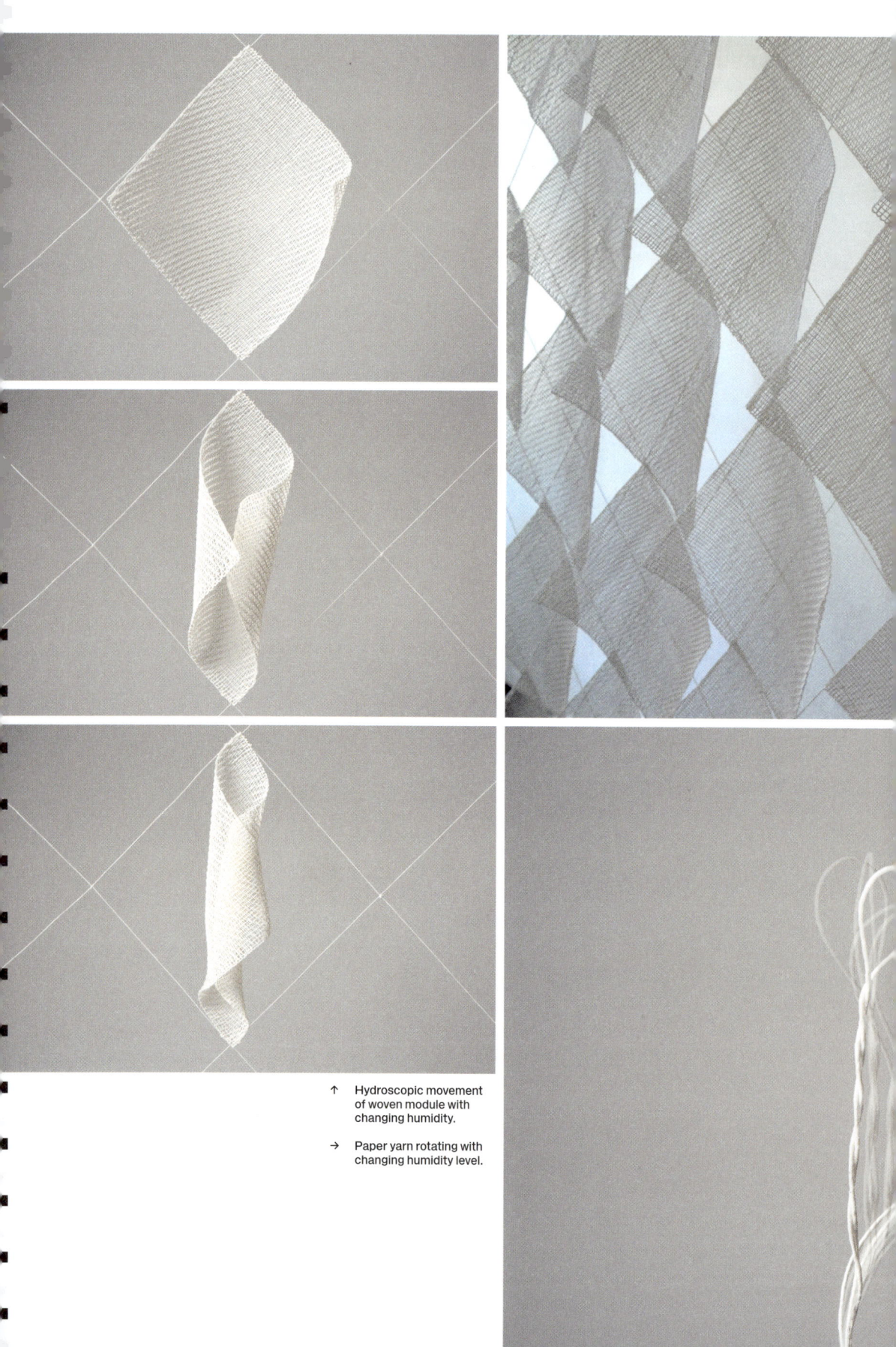

↑ Hydroscopic movement
   of woven module with
   changing humidity.

→ Paper yarn rotating with
   changing humidity level.

Hydroweave by Stefanie Eichler and Juni Neyenhuys is a reactive textile that works on the principle of "hygroscopic motion" — the swelling and shrinking of natural fibers. Cellulosic paper yarn is woven into square modules that form a membrane. When exposed to moisture, the cellulose fiber swells and induces a rotating yarn movement, which serves as a natural material actor that changes the overall shape of the textile modules. The diagonal twill-weave causes the modules to fold and roll on their free ends. Through punctual fixing to a supporting net structure, the movement and openness factor of the overall surface can be controlled. Further textile parameters that influence the movement and "program" the fabric performance are yarn diameter, twist direction, thread density, weave pattern, and module proportion. Depending on the direction of textile construction, the surface can either open or close with increasing humidity and it returns to its initial state when dried.

In addition to its functionality, the system also acquires an aesthetic-sculptural character. *Hydroweave* showcases a material property that is commonly considered problematic: the shape-change of cellulosic material, such as paper, under the influence of moisture. The result is an adaptive membrane that responds to the environment without any technical actuators or sensors. In addition, the modules are made entirely from biodegradable and renewable natural material.

Hydroweave thus combines function and sustainability in a unique way: an everyday material — paper — is used according to its inner properties and therefore unfolds a completely new design potential. With water vapor, the modules curve or stretch depending on the setting, and can thus create surfaces that open or close adaptively, to ventilate or separate areas, example.

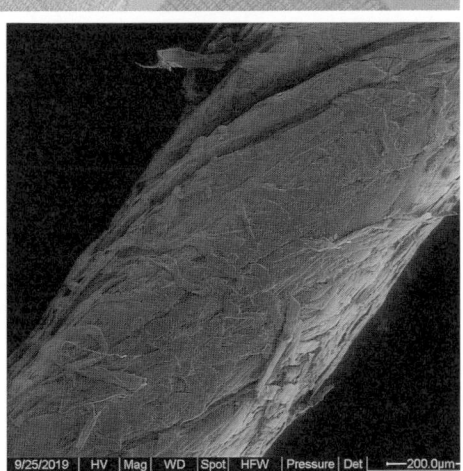

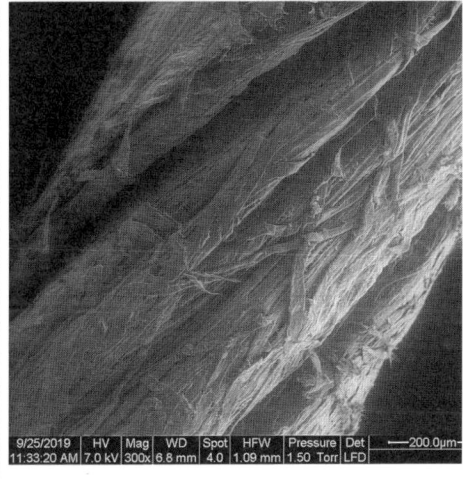

↑↑ Microscopic scan of dry paper yarn.

↑ Microscopic scan of humid paper yarn.

Skeletal Surface

Team

Xingwen Pan, Weißensee School of Art and Design Berlin, Department of Textile and Surface Design
In collaboration with Cluster of Excellence *Matters of Activity. Image Space Material*, Humboldt-Universität zu Berlin and Max Planck Institute of Colloids and Interfaces Potsdam/Golm, Department of Biomaterials
Mentoring team: Prof. Christiane Sauer, Ebba Fransén Waldhör, Maxie Schneider, Dr. Lorenzo Guiducci, Dr. Khashayar Razghandi

Context

Design Prototype, MoA Design Research Studio, *Scaling Nature (X) — Extended*, 2020

Material

Pre-stretched Jersey fabric, hot-melt glue, hand-drawn patterns with glue gun

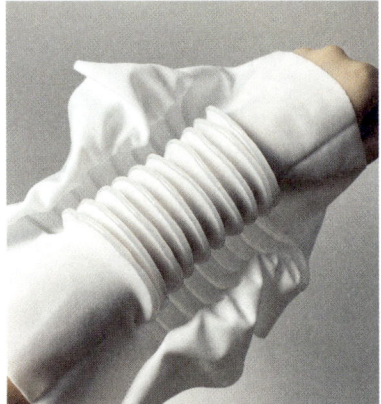

In contrast to *The Flip*, which used concentric lines, this project focuses on the effect of linear patterns shaping the three-dimensional fabric. The glue line layout and thickness become design parameters for the resulting shapes: by drawing the same pattern with 1 mm or 3 mm glue thicknesses, the final structures behave differently: a 1 mm glue line is strongly affected by the elasticity of the fabric, resulting in flat wavy structures, while a 3 mm glue line remains stiff enough to generate large volumetric structures.

By repetition of simple parallel patterns and by combination of gluing from front and back, a three-dimensional surface with alternating convex and concave rhythms is formed. The bending reinforcements not only give shape but also exhibit strong elastic behavior that can be used to fix the fabric, for instance onto parts of the body. This opens up possible applications in the context of clothing and fashion design. The delicate objects turn into textile architectures that are animated by effects of light and shadow, reflection and translucency.

SKELETAL SURFACE

This project is based on Xingwen Pan's design research project *The Flip* developed in *Scaling Nature (1) Wrinkles* (see p. 78), which transferred the flipping motion of a mushroom cap for releasing spores into a design for textile shaping.

Skeletal Surface explores how to combine the flexibility of elastic fabric and the stability of rods to transform a pliant two-dimensional surface into a semi-rigid three-dimensional object. Lycra fabric is stretched to its maximum over a metal frame and then different patterns are drawn by hand onto it with a hot-melt glue gun. After the glue has solidified, the tension of the fabric is released. The geometric pattern of the solidified glue partially inhibits the elasticity of the flexible fabric. As a result, the fabric shapes into three-dimensional convex and concave structures that result from the interplay of material forces. In this way, the soft fabric is tensioned along a "skeletal" support structure.

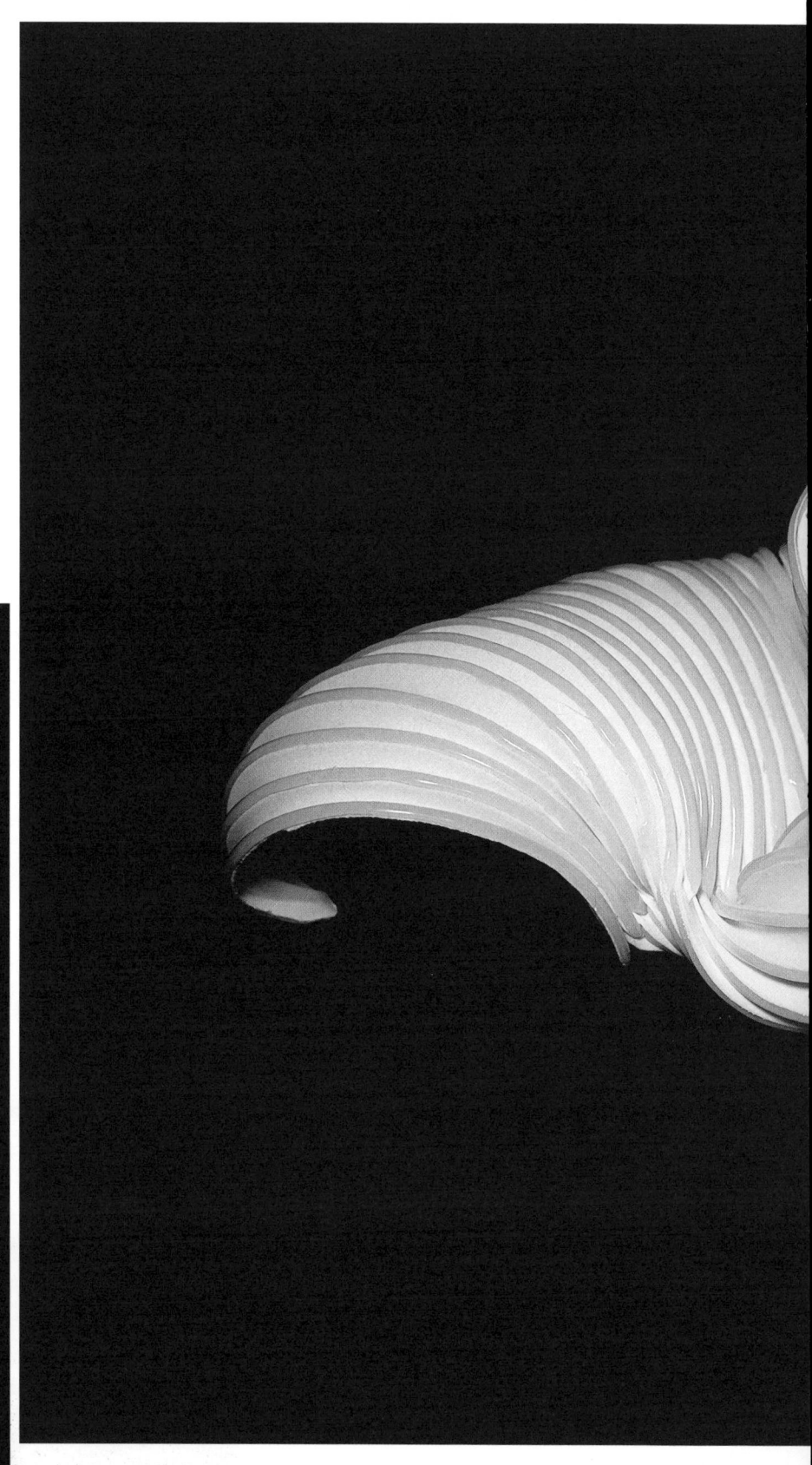

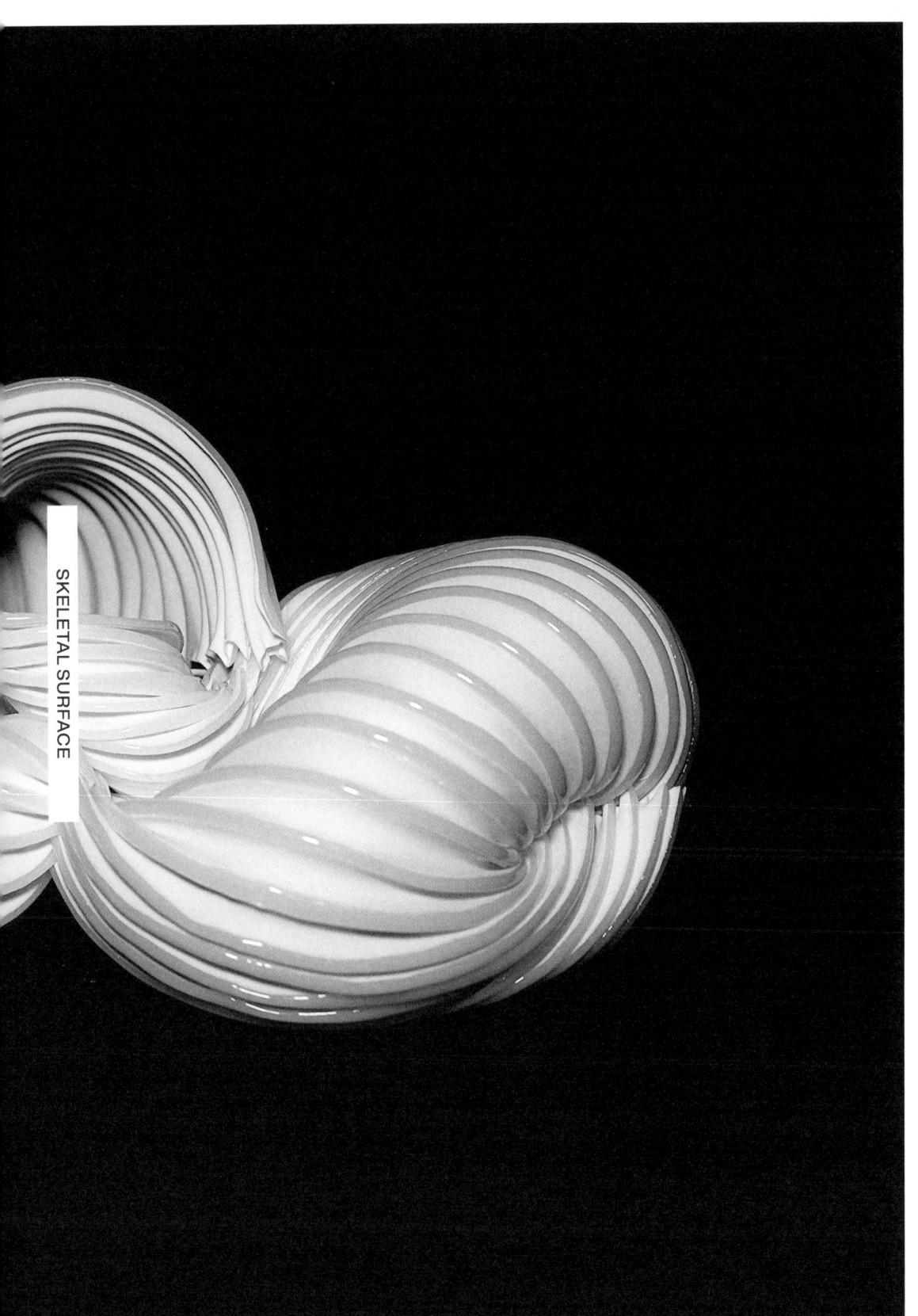

SKELETAL SURFACE

Weaving Wrinkles

Team

Samira Akhavan, Weißensee School of Art and Design Berlin, Department of Textile and Surface Design
Textile Prototyping workshop *Weaving Wrinkles* by Ebba Fransén Waldhör

Context

Design prototype as part of MoA Design Research Studio, *Scaling Nature (1): Wrinkles*, 2019

Material

Linen, cotton, and elastane yarn, single- and double-layered weave in various patterns, woven on a TC-2 Jacquard loom

A common objective in weaving is finding a balance between material and structure so that the resulting fabric is flat and smooth. In the workshop *Weaving Wrinkles*, participants experimented with the opposite — creating three-dimensional and highly textured fabric samples by intensifying contrasts. The interplay between different materials and yarn structures was explicitly designed to force the fabric out of its two-dimensional plane.

For this, the participants had to rethink weaving in terms of elasticity and stiffness, tension and release. Wrinkles and folds were created by combining elastic and stiff yarns in one fabric, utilizing the fact that the yarns are under tension during the weaving process. When taken off the loom, the elastic yarns contract, which forces the stiff material to wrinkle, bulge, and fold. The samples were made on a computer-controlled Jacquard loom that is conceived especially for rapid prototyping in the field of textiles. Each warp thread is individually controllable, which allows for complex and non-repetitive patterns — yet the weft is inserted manually for freedom in material choice.

Samira Akhavan experimented with combining different yarns in a weave that have both single and double layers in the same pattern. For double-layered weaves, the vertically arranged warp threads are divided into sections that are woven in an alternating rhythm, splitting the fabric into parallel, separate planes. Multi-layer planes automatically have less thread density than a single layer, since the given number of warp threads is split between the layers. Less density and more space between the warp threads gives the individual yarns more room to move — the interlocking is looser — and the whole fabric twists in unexpected ways. By combining linen — as a stiff yarn — with the highly flexible elastane in single and multi-layer bindings, a variety of fascinating samples were produced that explored the effect of different binding patterns for shaping the overall fabric.

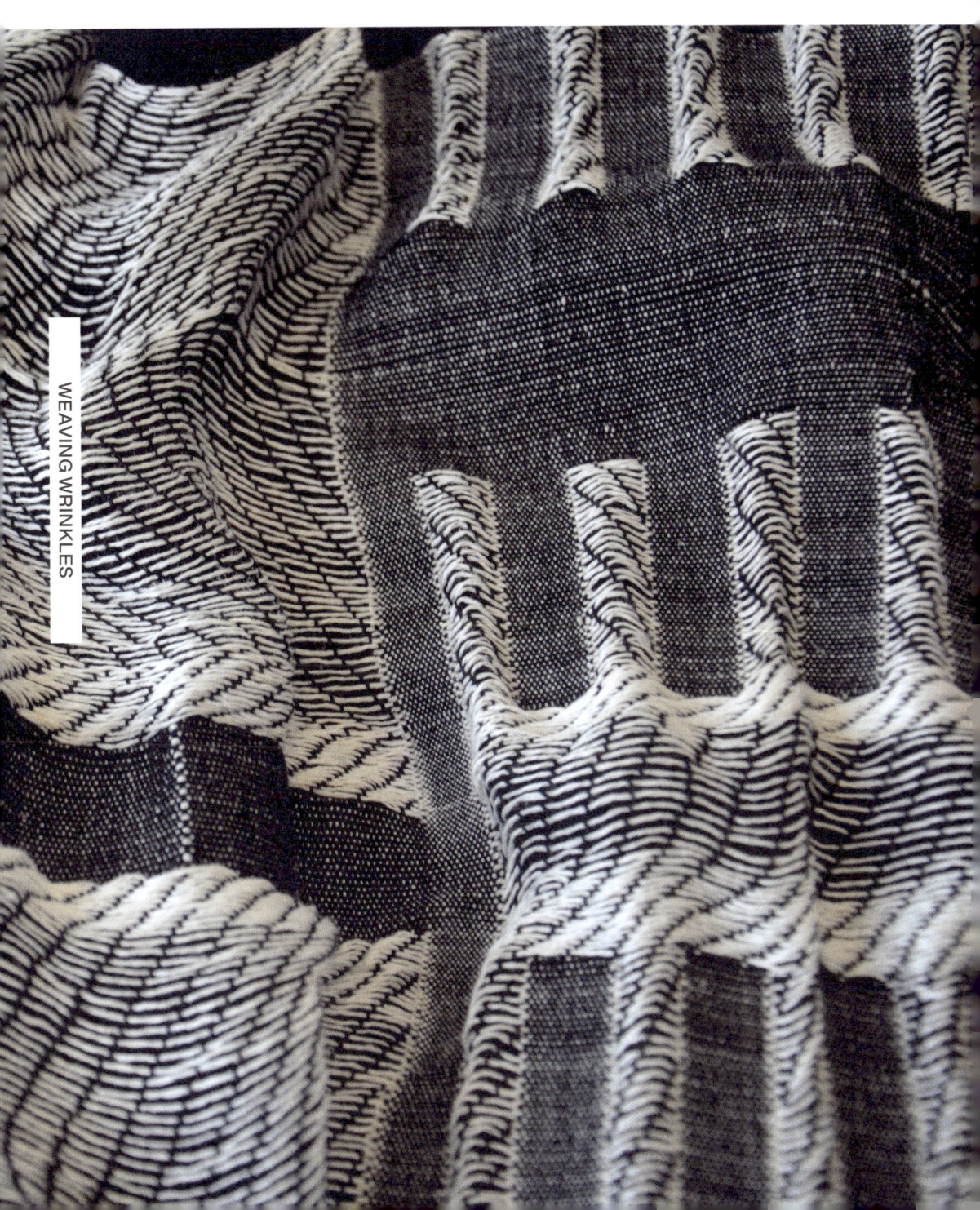

WEAVING WRINKLES

Living Beings

Team

Nelli Singer, Weißensee School of Art and Design Berlin, Department of Textile and Surface Design, Master Studio Prof. Christiane Sauer
In collaboration with Dr. Michaela Eder, Max Planck Institute of Colloids and Interfaces, Department of Biomaterials Potsdam/Golm

Context

Design Prototypes, Master Thesis, 2020

Material

Wood veneer filaments of different width and length, woven and knitted

↓ Double-woven structure as adaptive surface, made from cotton with integrated wooden filaments.

↓ Movement of veneer strips dependent on grain direction.

In nature, plant movement can be triggered by environmental stimuli such as temperature or humidity. The project *Living Beings* explores how to transfer such cellulosic shape change into a textile design concept. The cell walls of wood consist of cellulose microfibrils that swell with humidity. This causes the plant structure to move, twist or bend in relation to the angle and orientation of the microfibrils. Nelli Singer utilized this principle to animate wooden filaments cut as thin strips from veneer. The angle of the cut-out in relation to the grain direction determines the kind of movement. By working the threads into a woven or knitted textile construction, their behavior can be further "programmed" as a coherent material system: textile binding, pattern, number, and width of threads or size of loops were the parameters for programming through experimental design prototyping.

For the woven samples, wooden filaments serve either as warp or as weft yarns for a plain weave. Elements have to be bendable in warp direction when being processed and rolled up on a loom. For this purpose, strips of ash wood were cut at right angles to the grain and fixed on the warp cotton threads. The resulting fabric samples change shape with humidity and bend back to their flat state upon drying as the cotton yarns guide the wooden strips back to their initial position in the weave. For the use as weft elements, millimeter-thin compression wood or paper-laminated ash-wood strips with a perpendicular or diagonal grain were cut. The specific wood types influence the bending properties of the textile. Compression wood strips, gained from strained areas of a tree, can bend strongly without breaking. Even a slight water spray leads to a strong and fast movement. Further, the interaction of two layers with each other was explored in a cotton double weave with integrated wood filaments in both layers — warp and weft. This caused a spatial, nearly architectural configuration of the textile when activated.

The knit experiments are constructed from wooden filaments only. The thin strips are initially pre-formed into loops by wet-form bending. The resulting dried wavy elements are interlocked by hand into looping patterns. When wetted again, the wavy modules exhibit a shape-memory behavior with their loops opening up and expanding. Stitch size, stitch number, and arrangement also determine the way the single strips interact: some wooden surfaces bulge while others elongate depending on their textile composition and module arrangement. With this interlocking knit technique, comparatively stable structures were formed into small objects or assembled into room-scaled surfaces.

The designer describes the fascinating nature of her *Living Beings* objects as follows: "Due to the material's inner activity generating the movement, at first their movements appear inexplicable, almost like virtual animations, even though they are based on purely natural processes. When viewed in motion, the resulting textile surfaces evoke the impression of liveliness".

← Hygroscopic module, knitted 4 mm maple strip, cut perpendicular to grain direction, right-left stitches, stitch size 10 mm, size: 15 × 5 × 2 cm, expansion and contraction along length by 40% in approximately 5.5 wminutes.

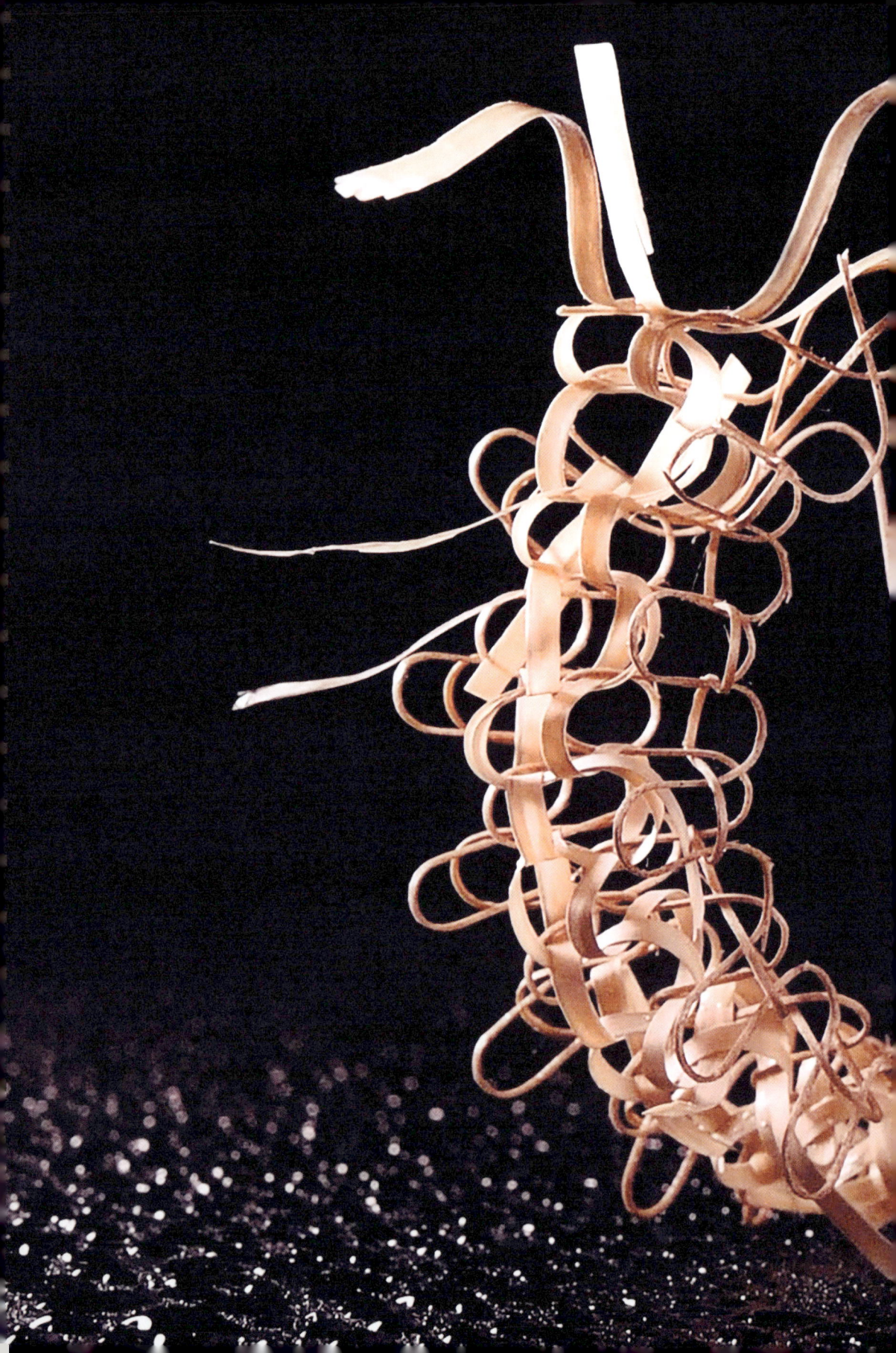

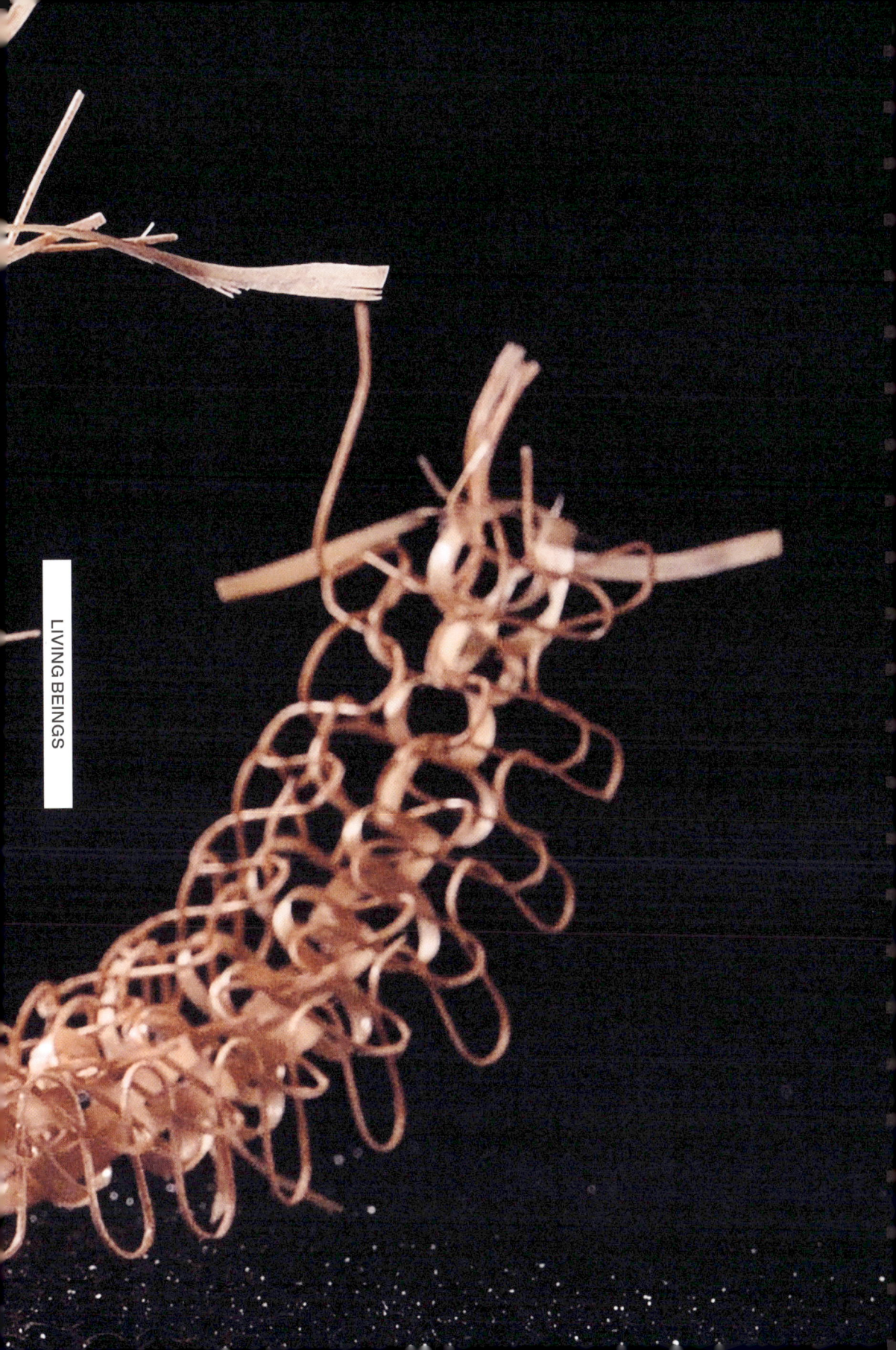

→ Hygroscopic module array, knitted 1 mm strip, maple veneer cut in grain direction, stitch size 10 mm, alternating 11 right stitches and 11 left stitches, size: 118 × 4 × 4 cm.

← Hygroscopic curtain, knitted 1 mm strip, maple veneer cut in grain direction, stitch size 10 mm and 20 mm, size: 182 × 126 × 4 cm.

Technical Textiles Between Function and Design—A conversation with Heike Illing-Günther

CHRISTIANE SAUER:
You have been working on the development of technical textiles formerly as Research Director and now as Managing Director of the Saxon Textile Research Institute (STFI). The book title *Architectures of Weaving* indicates the idea of a textile as a three-dimensional construction. What's your perspective given your background in textile technology?

HEIKE ILLING-GÜNTHER:
Textile is always three-dimensional for a textile engineer. Only mathematical simulations abstract it into one or two dimensions. Even membranes or nonwovens have this third dimension: fibers are layered to become, for example, a filter.

CS:
Yet I think textiles are considered mostly as flat membranes or decorative add-ons, not as potential technical elements in architecture. We can implement electric circuits or even light-emitting fibers in textile surfaces. Your curtain could become your lamp and even the switch for it. Textiles can also generate power through piezo fibers, photovoltaic coatings, or they can absorb sound and insulate against heat and cold. Technical textiles are also ubiquitous in everyday life: from nonwoven or woven tea bags to 3D warp-knitted mattresses and glass or carbon-fiber reinforcements in concrete building. Most people, though, aren't aware that these products are based on textile technology and might never have come across the term "technical textiles." Why is this? Are technical textiles hidden too well in our daily lives?

HIG:
A technical textile is first of all characterized by its function, not by its design. Hardly anybody knows that textiles are found in the body of trucks, because they are laminated and printed and therefore hidden and not recognized. Also, technical textiles are not always classic textile structures, but also tape structures, net structures, grid-like structures, that differ from the common understanding of textile that is associated with cloth and clothing.

CS:
Maybe technical textiles need more self-confidence to reveal themselves. Imagine the Eiffel Tower with a cover hiding its structure! It would never have gained its lasting popularity.

It's a purely technical structure, but at the same time it's designed with a fascinating sense of aesthetic that has been admired by generations. Its three-dimensional structure of steel trusses could even be interpreted as a kind of 3D "woven" meshwork. Also for technical textiles we need this kind of understanding of design. It's not about design *or* technology, but a merging of both.

I remember my first encounter with technical textiles at the *Techtextil* fair twenty years ago. I guess I was the only architect there and I was blown away by the aesthetics of these structures: delicate three-dimensionally woven translucent sandwich structures, shiny ultrathin aluminum-coated membranes and many more. The companies pointed out their properties, which are amazing, but no-one spoke about their fascinating aesthetic and design potentials. Since technical textiles are produced as a material for a specific purpose, the designing of both function and aesthetic is possible from scratch. This bears a lot of potential for visible applications of these textiles. In the mutual cooperation between our institutions, the STFI-Saxon Textile Research Institute and the Weißensee School of Art and Design, we experiment with this approach when designing with spacer fabrics (Fig. 1), fiber recycling (Fig. 2), or *Kemafil*® technology (see Case Study *Scaling Fiber* p. 149). But is design and aesthetics also addressed in your daily industrial commissions and research?

HIG:
We recently had a project for an automotive interior where the coarse fiber structure of the dashboard was left visible as a surface. You can surely save material and make things lighter when designing without adding further layers. I'm not sure if it will work out because at the end of the day the question is whether the consumer will accept it. The growing awareness and movement towards recycling and reuse will support the increasing acceptance of this aesthetic.

CS:
Do political and societal developments reflect in your research topics and developments? Textile is such a versatile material that inhabits all areas of our daily lives.

HIG:
Yes, we're always affected by political decisions and by trends in society. Actually, it's not possible for us to write a five-year research plan, because we don't know today what will be needed by the small and medium-sized companies — the main field of cooperation partners — in three years' time.

Textile technology can respond quite fast to novel requirements. Also, the topics for funding programs come up according to new political aims. As we're in close contact with industry, we're able to bring ideas and funding together

Fig. 1
→ *Wunderkammer.*
Spacer fabric design with
inner-grown crystals.
Madeleine Marquardt,
Weißensee School of Art and
Design, Design Studio Prof.
Christiane Sauer, 2016.

Fig. 2
↓ *New Blue.*
Textile fabric made from
recycled jeans fiber. Tim van
der Loo, Studio A New Kind of
Blue, start-up initiated from
Master Thesis at Weißensee
School of Art and Design,
Master Studio Prof. Dr. Zane
Berzina, 2019.

TECHNICAL TEXTILES

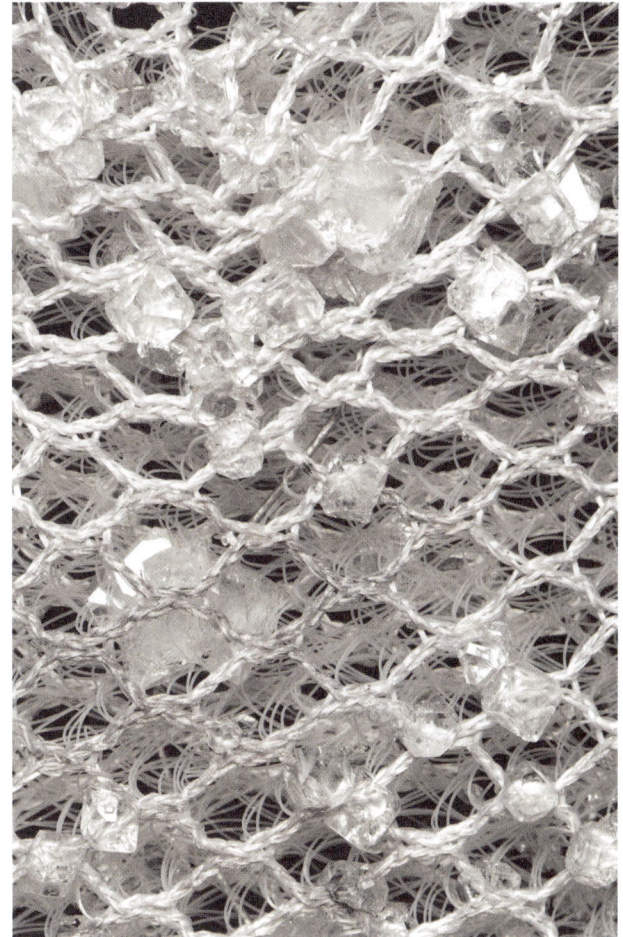

Fig. 3
↑   Geotextiles using
    Kemafil® coarse
    rope structures as
    reinforcement for
    landscaping to prevent
    erosion in road banks.

and react quickly to such upcoming tendencies. I just got a call from a colleague who has an idea for new battery systems based on nonwoven membranes. Now we're planning a project together, but a month ago I had no idea that I would ever deal with battery systems. We're now seeing great potential in this idea and want to learn more and develop new fields for application. So with technical textiles we constantly come in contact with completely different branches. Possibilities are endless once you start to discuss across disciplines.

CS:
Your institute specializes in nonwovens—a term that might be known only to experts. As nonwovens are used for filtering, this has gained enormous importance in the last pandemic years. How did this reflect in your production?

HIG:
For our Chemnitz institute, nonwovens had always been important, even before. The basic spun-bound and melt-blown technologies were invented and patented in the 1970s by our former GDR-colleagues at FIFT—Research Institute for Textile Technology—resulting in the nonwoven material that is now ubiquitous in the FFP2 protection masks that we wear every day. In 2020, we transferred from experimental to production mode in a short time in order to meet the soaring demand with albeit small production volume. After all, around 4.2 tons of nonwovens were delivered every week by mid-2020, a quantity sufficient for the production of at least 750,000

masks. With this challenge, our employees have gained valuable experience in terms of process optimization. Additional knowledge and new ideas for future research topics were generated.

CS:
STFI is one of the leading institutions of all the sixteen textile-research institutes in Germany. Is there a textile specialization in your region of Saxony?

HIG:
Twenty-five years ago, after the German reunification, the East German institutes were re-founded and took on special niches. The Chemnitz Textile Institute (FIFT) merged with the Research Center of Technical Textiles (WTZ Technische Textilien) in Dresden that specialized in nonwovens and coarse textile structures like ropes and nets. They brought with them the *Kemafil®* technology that was invented in the 1970s and allows the production of coarse rope structures. Since the 1990s, we've been using this technology to produce ultra-coarse filaments for geotextile applications. These are yarns with a diameter of 15–20 centimeters, which serve as reinforcements of unstable grounds and can prevent earth slides. We applied this technology in local projects, like in the area south of Leipzig where we have erosion problems in the former opencast mining area. After the reunification there was also a lot of road construction happening which needs artificial slopes. Due to steep angles, they tend to wash away with heavy rains. This became another application field for our ultra-coarse textile structures (Fig. 3).

CS:
How far do local textile traditions feed into your research? Would you consider the STFI a locally rooted institute?

HIG:
Yes, we're tightly connected to the local industry. Chemnitz is considered the Saxon Manchester, with a strong textile-machine building industry in the area—from big machinery to spare parts. We're in close contact with the local industry when discussing new ideas. For textile developments, there are no given calculation tables like in metal engineering. You rely a lot on people's experience. This needs exchange in person, so it is indeed a local and personal network.

CS:
Are there new developments you're working on in the field of architectural applications?

HIG:
We're currently discussing bringing hemp farming back to Saxony. Hemp is a really strong but also rough material; it's not quite suitable for clothing. We peel the hemp bark shortly after harvesting and get natural tape structures up to 4 meters long. They can be bound together to form grid structures or bio-composite rods. We're looking into using it as a reinforcement for wood

structures and as natural composites. (Fig. 4). Bio-based materials for architectural applications are a great future prospect. This idea isn't only interesting in terms of textile technology, but also with regard to society. Hemp is a modest plant that grows easily, which is ideal for the former surface-mining areas in the Lausitz area where there's still a high rate of unemployment. This textile idea could generate urgently needed jobs for the region that would suit the former local labor structures dedicated to workmanship and fabrication. We're currently talking to politicians to get further support for this project.

CS:
How do you react to current environmental challenges? Do you have an example from your practice?

HIG:
We look into materials to substitute plastic in products. For instance, fishing nets that are made from an outer layer of cellulosic fiber like lyocell with only an inner polymer core for tensile strength (Fig. 5). This reduces the amount of plastic in fishing nets and avoids spreading microplastic through mechanical abrasion of the outer layers. These are small steps, but each step is needed to reduce the load of microplastic in the environment.

CS:
How do you address the idea of recycling? Textile industry is one of the main factors of global waste streams. Even if garments cause the largest amount of textile waste, technical textiles might be more tricky to recycle: for instance, fiber-reinforced plastics, where carbon or glass fibers are bound into a duroplast matrix, can hardly be separated after use.

HIG:
The development in this field is very much related to the material's market value. High-tech fiber, like carbon, has a high market value even in its second cycle. Techniques are available to remove the resin and make the fibers available again. This is a starting point for textile recycling—also for carbon-fiber material. Even if the fiber loses some length, which means a loss in quality, it can still take several product routes. For the resin, duroplasts like epoxy can't be landfilled and incineration wouldn't be smart. Purely bio-based matrix materials are not yet available in the necessary quality and quantity; more research is still needed. And bio-based polymers aren't necessarily the better solution. For instance, sugar cane can be used to produce a bio-based Polyethylene (PE), which has the same chemical structure as petrol-based Polyethylene. This means that even the sugar-cane-based PE produces durable microplastic to the same extent as the conventional polymer. But Polyethylene can easily be recycled into new products because you can melt it again and

Fig. 4
↑ Hemp tape structures.

↗ Binding of hemp into a surface.

→ Hemp bands as reinforcement for wooden construction.

Fig. 5
↓ Warp knitting machine for netting structures.

again, but it has to be collected and sorted from other plastics. So one has to consider the whole cycle with all its implications.

CS:
This implies a shift from product design to process design, if we consider the whole cycle of production.

HIG:
Yes, we used to do a lot of down-cycling with textiles. Cheap and low-quality recycled materials were produced for use as simple products like cleaning cloths. Nowadays the question is addressed in more detail. Sorting textiles ahead is important before bringing it back in the loop. If we want to recycle a very simple white cotton T-shirt, it's an immense challenge to recycle it into a new white cotton T-shirt. Because the yarn fibers get shorter in the process, we can't produce the same white T-shirt from it. So why not produce T-shirt flakes and recycle them as a totally different but high-value product like Tim van der Loo's *New Blue* garment from recycled jeans fibers (Fig. 2), a synthesis between technology and design. This project was developed in a collaboration between our institute and the Weißenseee School of Art and Design. We should discuss, especially with the young generation of designers and students, how to rethink processes of assembly and disassembly and to generate new product ideas from this. But the consumer needs to have an awareness that every product has an impact on our lives and the environment. This should be taken into account more in society.

CS:
Where do you see an important field of future developments?

HIG:
I see the new research perspective coming up in energy management, CO_2 footprints and life-cycle assessments. As we closely cooperate with industry, their challenges are to simulate, model, and calculate this ahead of time. Together, we're still learning in this field.

CloudFisher®

Team

Aqualonis GmbH and WaterFoundation, Munich, Germany, in cooperation with the local community of Sidi Ifni, Morocco

Context

World's currently largest fog collector facility, developed by Aqualonis GmbH on behalf of the German WaterFoundation and funded by the German Federal Ministry for Economic Cooperation and Development (BMZ) and the Munich Re Foundation, 2013–18

Material

3D warp-knitted spacer fabric (PES), supporting triangulated net (HDPE), fixings, metal frame, drainage system

On the heights of Mount Boutmezguida (1,225 metres) in the Anti-Atlas mountains in Morocco, unusual textile structures rise above the clouds that develop from the nearby Atlantic Ocean. They are fog collectors, which catch tiny droplets of water and guide them into pipes to fill the cisterns of the Berber natives who settle at the foot of the mountain.

Fog-collecting with fishing nets or simple textiles has been practiced since the 1960s, when Carlos Espinoza Arancibia first introduced this pioneering invention in Latin America.[1] But the harsh environment places high demands on textiles. In order for the water harvest to succeed, it needs a membrane system of high tensile strength to withstand storms and good UV-resistance in extreme sunlight. For the WaterFoundation, Peter Trautwein developed the concept further to support regions suffering from water scarcity in order to enhance local agriculture and development. With the support of a team of researchers from the Department of Ecoclimatology,

Technical University Munich, they developed a resilient and low-maintenance fog-catcher system that is able to collect high-quality drinking water from fog.

The catcher consists of a UV-resistant multi-layer screen construction with a 3D spacer fabric that catches the fog and a robust triangulated netting that protects it from the mechanical impact of wind. The custom-made spacer textiles combine high air permeability with a large surface area provided by the three-dimensional structure of the entangled polymer filaments. This improves the water condensation and generates a high yield. The textile layers are elastically mounted onto metal frames to withstand high wind speeds during storms.

The droplets caught by the installed 31 Cloud-Fishers merge into larger drops that then fall into a gutter, from where they flow via pipes to different reservoirs. An average of twenty-two liters per square meter is harvested annually. With an overall installation of 1,600 square meters, that makes about 35,000 liters per fog day and a constant average of twenty-two liters per day per villager. The inhabitants of rural Sidi Ifni are very proud of their misty water source. Especially in remote areas, drinking water is conventionally carried on long walks from distant mountain springs — a task often undertaken by the youngest.

In order to prevent desertification of dry but misty regions across the globe, the WaterFoundation continuously researches possible regions for application and has made information on the systems publicly available, to support local initiatives with training and implementation. To effectively place the catcher in the mountain terrain, a preliminary study is carried out with the support of local communities to understand meteorological data such as the frequency of fog, wind speed and direction, relative humidity and temperature, precipitation, and accumulated water amounts.

1   C. Espinoza Arancibia, *Making Use*, accessed April 17, 2022, https://makinguse.artmuseum.pl/en/carlos-espinosa-arancibia/.

Adaptex

Team

Prof. Christiane Sauer, Ebba Fransén Waldhör, Maxie Schneider, Weißensee School of Art and Design Berlin, Research Group DXM - Design Experiment Material
Paul-Rouven Denz, Puttakhun Vongsingha, Natchai Suwannapruk, Dr. Jens Böke, Priedemann Facade-Lab GmbH, Großbeeren
In collaboration with: Fraunhofer Institute for Machine Tools and Forming Technology IWU, Carl Stahl ARC GmbH i-Mesh, Numana Italy, ITP GmbH, Schütz Goldschmidt Schneider Ingenieurdienstleistungen im Bauwesen GmbH, Verseidag-Indutex GmbH

Context

Research prototype. Research project funded by the German Ministry of Education and Research BMBF funding guideline Zwanzig20 – Partnership for Innovation, within the framework of innovation network smarthoch3, 2017-22

Material

Shape Memory Alloy (SMA) wires
Wave: Stainless-steel nets, laminated glass-fiber textiles
Mesh: Glass-Fiber rovings, basalt-fiber rovings coated with PVDF

Biological processes, such as the skin's ability to react to temperature, have long inspired the idea of adaptivity in architecture. Recent findings in smart materials drive the development of adaptive and kinetic architectural surfaces. By integrating smart material components into textile structures, thermal conditioning can be provided – without the use of electricity.

Shape memory alloys (SMAs) are smart metals that react to temperature by changing shape. As they heat up they contract by a few percentages. In the form of thin metal wires, they are quite strong relative to their weight and diameter and have mostly been used as actors in technical devices. They require less space than motorized control, operate silently without wear or corrosion, and are durable – with up to more than 100,000 motion cycles. Until now, the use of SMA for architectural applications has been very limited, since they were designed for the scale, precision and standards of mechanical engineering.

Adaptex interprets the SMA wire as filament for a textile structure. This enables the transfer of their performance onto a building scale, since textiles can be produced in endless lengths that can cover whole facades. They can also easily be mounted for energetic retrofitting on existing buildings.

The two systems *Adaptex Mesh* and *Adaptex Wave* both use the ambient temperature to enable the closing and opening of their textile structure. The adaptive shading prevents overheating of buildings and urban spaces through temperature-driven material actuation.

↓   *Adaptex Wave*, open.

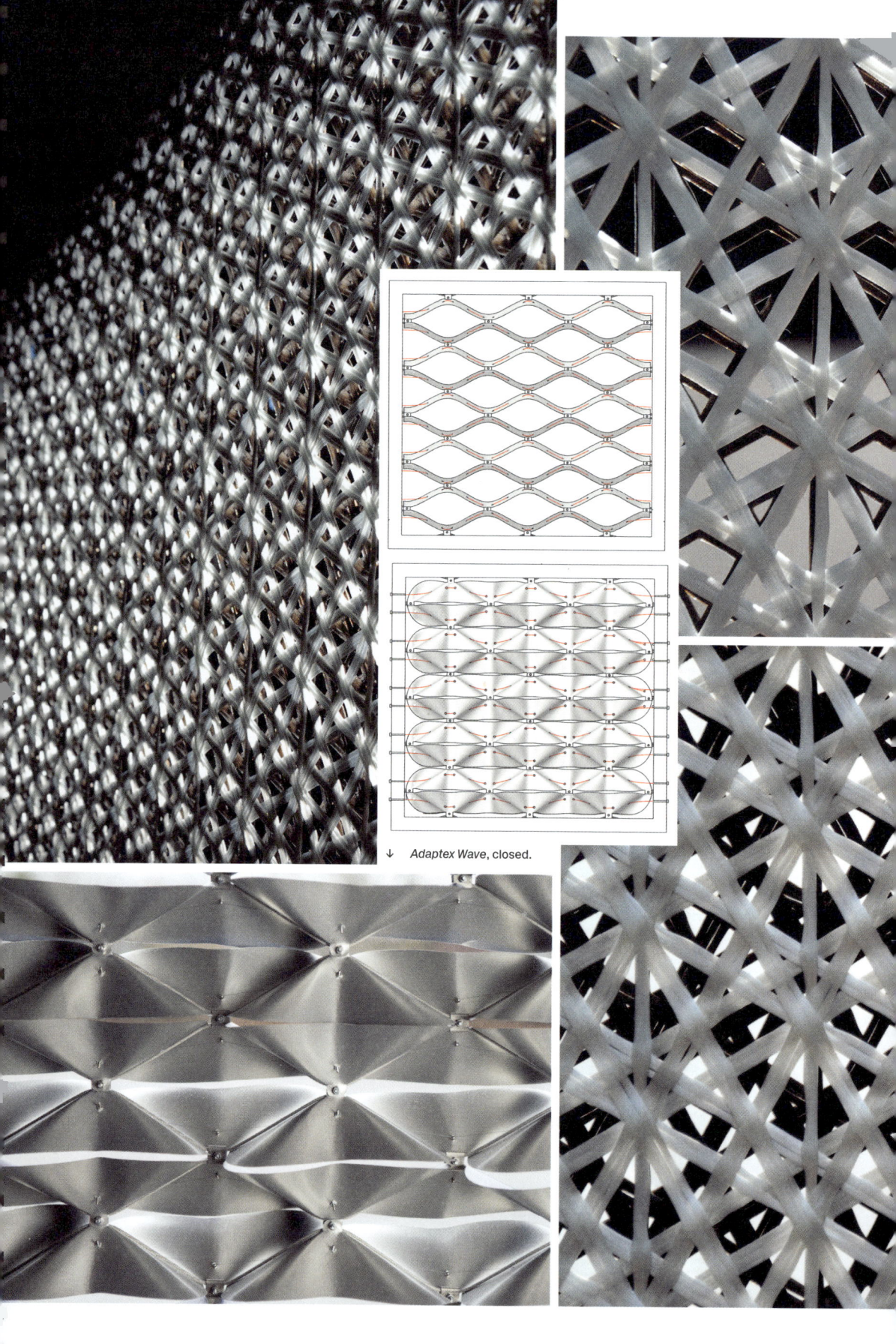

↓   *Adaptex Wave, closed.*

Adaptex Mesh consists of two textile layers with identical patterns placed in front of each other. When activated, the vertically integrated SMA contracts by 3% and allows one of the layers to slide upward, reducing the overall permeability of the surface. The multiaxial arrangement of glass- and basalt-fiber filaments balance the force of the integrated SMA and prevent fabric deformation. When the temperature drops and the SMA cools, the textile screen returns it to its initial position through gravity alone.

Adaptex Wave utilizes textile bands attached to a stainless-steel cable net at specific points, creating a three-dimensional geometry and strain in the textile. The tailored glass-fiber-laminated bands are interwoven with a linear SMA wire. When exposed to intense sunshine and the resulting rise of temperature, the SMA wire contracts, causing the bands to buckle and close. Due to their inner material strain, the textile bands move back to their initial position when the SMA wires relax with dropping temperature.

Because the SMA wires respond to tempera-ture, local climate data analysis plays a central role in determining the position of the *Adaptex* screen. It can be mounted on the facade exterior or inside the cavity of double-layered glazing. For additional user comfort and more independent control, it is also possible to activate the SMA through short electrical impulses.

In the journey towards a resource- and climate-friendly architecture, solar shading will play a central role to meet the impact of global warming in reducing the energy input for building climatization and protect interiors from overheating. This is exactly where the self-sustaining temperature-operated tech-nology of *Adaptex* comes into play.

↑ *Adaptex Mesh*, open.

↓ *Adaptex Mesh*, closed.

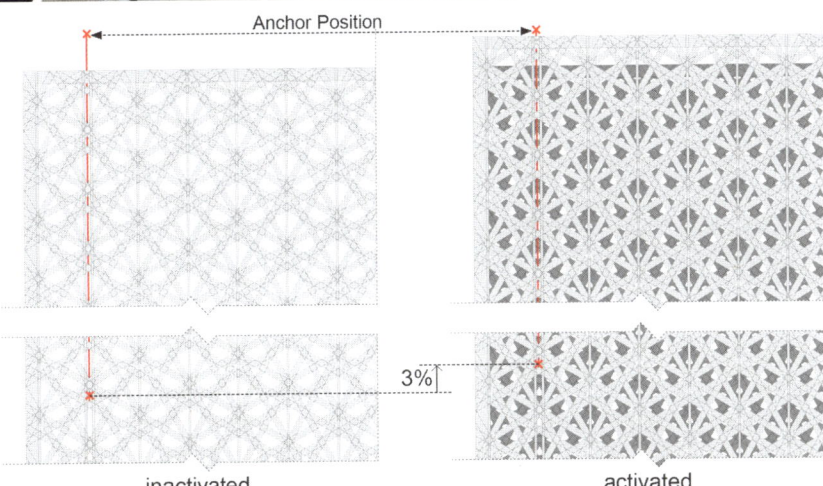

Anchor Position

3%

inactivated

activated

Black Hair Tents—Dealing with the Environment through Absorption Instead of Repellence

Kristina Pfeifer

Fig. 1
Members of the families
Şurgum and Şimşek in front
of their Yörük Tent in Western
Turkey, 2005.

ANCIENT ORIGINS

Although they have played a significant part in human history, facilitating human survival and the building of civilizations, black hair tents are little-known by the public. This type of tensile mobile architecture is used by nomad tribes of Northern Africa, the Middle East, and the Tibetan Plateau. Their awnings bear the tensile load of the construction and consist of coarsely woven hair, usually black goat hair, but with slight variations.[1] It is assumed that these structures originated in regions of the Arabian Peninsula and beyond, but since their construction is based on mostly decomposing materials, archaeological records are scarce.[2] Surprisingly, the oldest evidence is a description in Exodus 26: 7–14

---

[1] Variations include other animal hair types like camel hair mixed with the goat hair, or the exclusive use of yak hair, as seen on the Tibetan Plateau, for example.

[2] R. Cribb, *Nomads in Archeology* (New York: Cambridge University Press, 1991), 66.

Fig. 2
Newly pitched roof of a Yörük
tent in Western Turkey, 2007.

in the Hebrew Bible.[3] The text provides guidance for installing goat hair cloths for the tent over the tabernacle (fifteenth to thirteenth century BCE). Moreover, archaeological findings of Assyrian stone reliefs (668–627 BCE) show section views of tents that were constructed in a similar way to contemporary Kurdish hair-tent types in Eastern Turkey.[4]

A SHELTER FROM LOCAL MEANS

The tensile-loaded part of the tent is also the "skin," while the pressure-load role is taken by wooden poles with ridge pieces or ridge bars and/or arches made from local trees or with wood traded from neighboring settlements. The tension of the awning is achieved by far-stretched guy ropes, fixed to pegs on the ground. The tent designs are well adapted to local needs, whether far-stretched or aerodynamically optimized for heat and sandstorms in hot desert regions, or compact with steeper roofs in the mountains.

3   T. Faegre, *Tents, Architecture of the Nomads* (New York: Anchor Press, 1979), 9–10.

4   H. Klengel, *Zwischen Zelt und Palast—Die Begegnung von Nomaden und Seßhaften im alten Vorderasien* (Leipzig: Koehler & Amelang, 1971), 102, 119.

Fig. 3
Roof of a Kurdish tent near
Van in Eastern Turkey, 2005.

Fig. 4
Eaves' fixation of a Kurdish
Tent near Van in Eastern
Turkey, 2005.

Fig. 5
Man and woman sewing a new
Yörük tent in Western Turkey,
2007.

With the domestication of particular animals well-equipped for environmental hardiness, for example the goat in the Middle East and Northern Africa, and the yak on the Tibetan Plateau, humans managed to inhabit regions where forage is scarce and sensitive. The black hair tent derives from these limited resources, being a highly optimized design that can be built from few means. The animal's hair is shorn once or twice a year, and then traditionally spun with drop spindles, among which the cross-stick whorl spindle is a specialty.[5] The single twisted yarns are then twined into a two-yarn-thread and woven into a coarse plain weave on portable looms.[6] Lastly, the woven strips are sewn together into the shape of the awning.

In the 1930s, settled nomads in Turkey specialized in the production of these weaves. And from about 1960 onwards, this development led to huge facilities with automated machines mass producing textile panels for Turkey, the Arabian Peninsula, and the North African states.[7] Nowadays, these woven textiles are not only used for traditional tents, but also for modern architectural constructions like festival halls or tourist pavilions. These contemporary experiments in implementing the ancient weave are a first step toward investigating the suitability of this ancient textile in modern designs. But what do we hope to gain from it? To answer this, we need to look at the traditional use of the tent.

5   H. Böhmer, *Nomaden in Anatolien: Begegnungen mit einer ausklingenden Kultur*
(Ganderkesee: REMHÖB-Verlag, 2004), 187, 118.
6   E. Broudy, *The Book of Looms* (Hanover and London: University Press of New England, 1979), 41.
7   K. Pfeifer, *Yörük Black Tent. Adaption in Design in the Course of Changes in Production* (Vienna: Technical University of Vienna, 2015), 191–97, 219, 263–67, 271.

When I first saw a black hair tent, its inhabitants explained to me that this tent is cooler than any other and effectively withstands rain and sandstorms. Baffled by their statements, I began my journey of investigation. With the next rainfall, I had the opportunity to experience the proclaimed impermeability to rain. I decided to test this phenomenon in a laboratory. The preliminary results indicated that such a tent leaked horribly. But how could that be? Humans had dwelled under them for millennia. Something must have been wrong with the laboratory conditions. It soon turned out that the assigned European laboratory standard setting (Beregnungsprüfung nach Bundesmann, Ref.-Nr. EN 29865: 1993) did not conform to the holistic nature of the tent. After further expert exchange and research on ancient building techniques, I learnt how a slight variation of the laboratory settings changed the outcome to a rain impermeability of 99%.[8] What had happened? The European standards had never considered a textile that deals with water not through surface repellence but through absorption of water. The awning of the black hair tent soaks up the rainwater and guides it within its plane to the low

8   K. Ambrosch (author's maiden name, now Kristina Pfeifer), *Karahane* (Vienna: Technical University of Vienna, 2005), 156–57, 166–67.

BLACK HAIR TENTS

Fig. 6
Weaver Adnan Yarar with his treadle loom producing black tent cloths in Western Turkey, 2007.

Fig. 7
Steam rising from a Yörük tent directly after rainfall in Hungary, 2006.

Fig. 8
Rain impermeability test with a
black goat hair weave of an old
Yörük tent in Vienna, 2004.

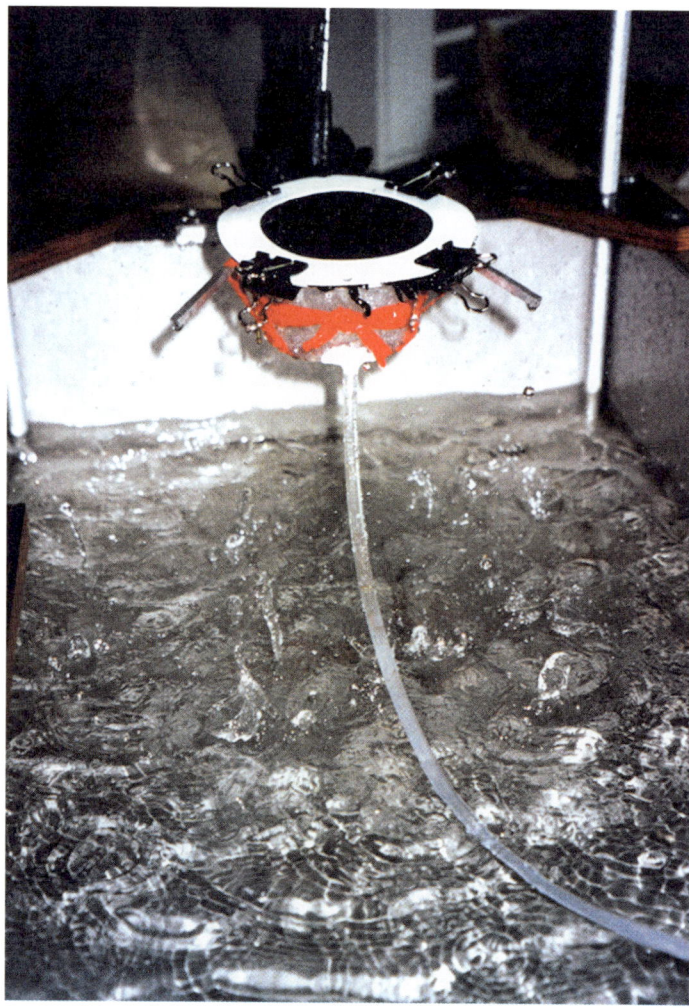

fringe, where it drips down to the irrigation trench on the ground. This indigenous textile simply integrates the unwanted water, thereby protecting the inhabitants from it.

While the impermeability to rain is not easy to understand, resistance against sandstorms is more obvious. The coarse weave is rough enough to withstand the grinding dust, and the far stretched guy-ropes and the many pegs under an aerodynamic awning sufficiently take the horizontal force and vertical suction effect induced by the wind. This, I experienced myself after a sandstorm in the Sahara Desert when the Berber tent remained unchanged while the modern canvas tent next to it was crushed (Fig. 9).

Although using a black awning, the interior space is described to be cooler in hot climates than any other tent design,[9] allegedly achieving temperatures of 18°C–30°C cooler than the outside.[10] I experienced this essential cooling effect, emphasized by nomads in interviews, in Turkey and Morocco during the hottest season. Scholars have observed how the black color reduces the glare of the sun, while the coarse pores of the textile still allow in sufficient sunlight for visibility.[11]

9   T. E. Lawrence, *Seven Pillars of Wisdom* (New York: Doubleday, Doran & Co., 1936), 279–80.

10   R. Qobrosi, *The adaptability of the bedouin tent in the hot dry climate of Jordan* (London: Architectural Association of Architecture School, 2013), 9.

11   P. Drew, *Tensile Architecture* (Boulder, Colorado: Westview Press, 1979), 42; Qobrosi, *The adaptability of the bedouin tent in the hot dry climate of Jordan*, 9.

Fig. 9
Moroccan Berber tent (left)
next to a collapsing tourist tent
(right) after a sandstorm in the
Sahara desert, 2009.

However, scientific investigations into this issue are scarce. One experiment by Shady Attila with a computer-simulated Bedouin tent based on field samples, produced results that do not highlight the proclaimed cooling feature.[12] Here too, experience indicates that laboratory conditions or computer models based on Western standards are often not applicable to indigenous designs for hot climates, as was concluded by Attila himself,[13] and thus the simulation provides grounds for further questions.

There is a theory that the coarse black weave of the tent absorbs the sunlight more than other colors, heating the textile and nearby air layers so effectively that the thermal currents rise through the cloth into the exterior before affecting the interior. Attila has suggested, and I agree with this assumption, that these currents even draw out further air from the inside, providing subtle and constant ventilation. If this is the case, one would expect to see evidence of a passive cooling effect by solar heat absorption and transformation. A picture taken from a tent roof after a sudden rainfall shows steam rising from the awning as soon as the sunbeams strike the textile. This picture hints that in-depth on-site investigations into thermal conditions need to be carried out (Fig. 7).

12   S. Attila, *Assessing the Thermal Performance of Bedouin Tents in Hot Climates* (Doha, Qatar: Qatar University, 2014), 1.
13   S. Attila, *Investigating the Impact of Different Thermal Comfort Models for Zero Energy Buildings in Hot Climates* (Qatar: ASHRAE Conference, 2014).

The advantages of black hair tents are fascinating and various: Protecting inhabitants from both sun and rain, they can be used in conditions where settled life is impossible due to extreme aridity or climate impacts; the components are either completely bio-degradable (goat hair, plant fibers, wood) or safe for the environment (iron, steel for hardware); the textile itself can be repaired and renewed, while the wood and iron can be reused.[14] The materials can be gathered in harsh conditions from the robust goat or yak, or in an environment with very few trees. While anthropology provides more and more documentation of ethnic groups utilizing black hair tents, technical investigations are still lagging behind, though showing much potential for novel approaches to material developments. Now, however, with the pressing need to find sustainable material solutions, there is great impetus to explore a textile technology that has supported humans from ancient times until today.

14   T. Rouhi, *Mobile Vernacular Architecture in Northwestern Frontiers of Iran* (Vienna: Technical University of Vienna, 2018), 125–33.

Fiber Structures

Constructing with fiber on an architectural scale puts the design and definition of "fiber" at the center of interest. As found in vernacular building traditions like constructions with bamboo stalks in Asia, meter-sized linear elements of centimeter-sized diameters can be interpreted as fibers for an up-scaling of textile construction methods. Such natural and resilient elements are suitable for scaffolds and even small architectures. The transfer from traditional bamboo applications to a contemporary large-spanning fiber-based construction is shown in one of the following case studies. Even living plants are traditionally used to create outdoor spaces when on support structures like pergolas. A building-sized structure constituted from growing trees forms a living architecture that is showcased as a resilient though slow means of sustainable construction.

Textile thinking questions the paradigm of structural stiffness and opens it to the idea of softness and flexibility. Stability in an architecture-scaled textile system can be achieved through different approaches shown in further examples: utilizing the inherent elastic energy of natural filaments (rattan or willow) allows for the geometrical bracing of a three-dimensional structural textile. On a larger scale, GFRP bending-active rods are introduced in a tensile knit to create a resilient yet soft textile tower balancing tension and compression and exploring the relation between structure and skin. Furthermore, load-bearing textile composite structures are created from reinforcing tensile glass-, carbon- or basalt-filaments with a binding matrix. The fiber layout relates to the structural flow of forces, resulting in lightweight fiber geometries.

Up-scaled architectural-sized yarns can form durable screens when textile sleeves are filled with concrete or they can stay reconfigurable as entangled walls when filled with loose materials like residue fibers, granules or even earth that enhance a room's climatic or acoustic conditions.

Vedana Restaurant

Team
Vo Trong Nghia, Nguyen Tat Dat, Nguyen Duc Trung, Tu Minh Dong, Nguyen Tan Thang
VTN Architects, Ho Chi Minh City, Vietnam

Context
Restaurant in a resort hotel, Cuc Phuong Commune, Vietnam, 2020

Material
Bamboo, rattan ropes, purlins, fern thatched roof (*Gleicheniaceae*) with wire netting

Vedana Restaurant is located on the edge of Cuc Phuong, Vietnam's oldest national park, home to a diverse biosphere attracting thousands of tourists every year. VTN Architects placed the restaurant in the center of a resort next to an artificial lake that helps humidify and cool the air. Due to the openness of the building, the interior gradually rises to the outside via semi-open spaces, creating a rich spatial experience for visitors. They can perceive the interior of the bamboo architecture and the mountains with the lake simultaneously.

Bamboo, traditionally used in Asian architecture for centuries as scaffolding and for building bridges, pavilions, houses, and other structures, is a lightweight, strong, and inexpensive material and ideal in this context. To meet the challenges of the twenty-first century, VTN Architects have worked mainly with two species of bamboo: the flexible Tam Vong (*Thyrsostachys oliveri gamble*) and the more robust Luong (*Dendrocalamus barbatus*). They developed a manufacturing process that allows for the production of standardized modules

with environmentally friendly traditional surface treatments such as soaking in mud and smoking for antiseptic treatment and durability.

VTN's famous bamboo projects, inspired by forests and trees, use an innovative technique with prefabricated bamboo frames that rise like a tree and branch out into a cross vault. To compensate for irregularities in the individual bamboo poles and create dynamic and flowing spaces, multiple poles are bound together with rattan into large macroscopic fiber bundles.

The restaurant's three-gabled circular roof, with an area of 1,050 square meters, consists of two superimposed ring-shaped roofs and a domed roof, each separated by light bands. The radius is about eighteen meters, while the structure is almost sixteen meters high, making it VTN Architects' tallest bamboo structure to date.

Plane Tree Cube

Team

Prof. Dr. Ferdinand Ludwig, Daniel Schönle, Jakob Rauscher, OLA — Office for Living Architecture, Stuttgart; Baubotanik[1] Research Group, TUM School of Engineering and Design, Technical University of Munich

Context

Building demonstrator, contribution to the *Landesgartenschau* 2012 in Nagold, Germany, 2012–ongoing

Material

Trees (London plane/*Platanus acerifolia*), inosculated tree trunks and branches, steel construction

Branches and trunks of trees are characterized by living tissues that enable growth and maintain vital functions. In *Baubotanik* projects, these biological processes are combined with methods of technical joining to merge the living and non-living elements into hybrid architectural structures. Thereby architect and tree become "co-designers" who together shape a "building," which is under constant transformation. It will never be "finished." Even if desired stages of development are reached sooner or later, the future outcome depends on circumstances and factors that — to a great extent — cannot be planned for.

This processual approach to architecture can be illustrated with the example of *Plane Tree Cube*, an experimental Baubotanik project that was realized in 2012 in the city of Nagold, Germany. Using the so-called plant-addition technique[2], the ten-meter cube was realized as a completely green structure, providing the

---

1   The German neologism *Baubotanik* (German *Bau* = building; *Botanik* = botany) was established by the Research group Baubotanik at the Institute for Architectural Theory IGMA, University of Stuttgart. See: F. Ludwig and O. Storz. "Baubotanik - Mit lebenden Pflanzen konstruieren." *Baumeister. Zeitschrift für Architektur* 11/2005 (2005): 72–75.

2   See e.g.: F. Ludwig and D. Schönle. "Growth as a Concept", in *Hortitecture. The Power of Architecture and Plants*, ed. A. Grüntuch-Ernst (Berlin: Jovis, 2018), 65–71.

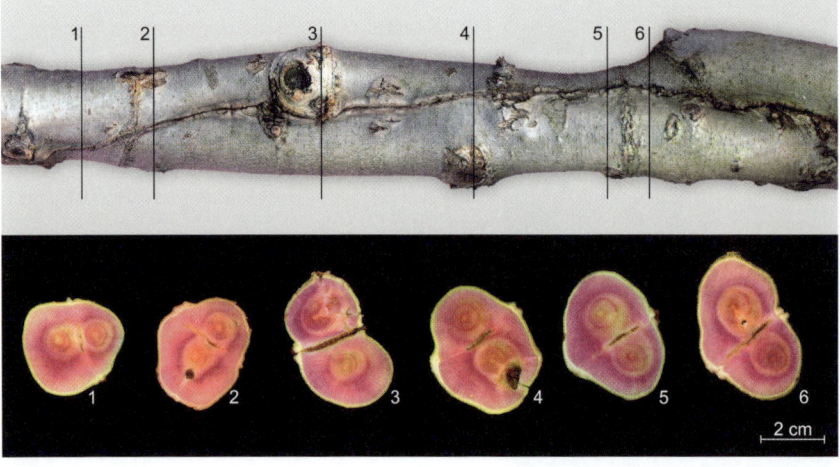

PLANE TREE CUBE

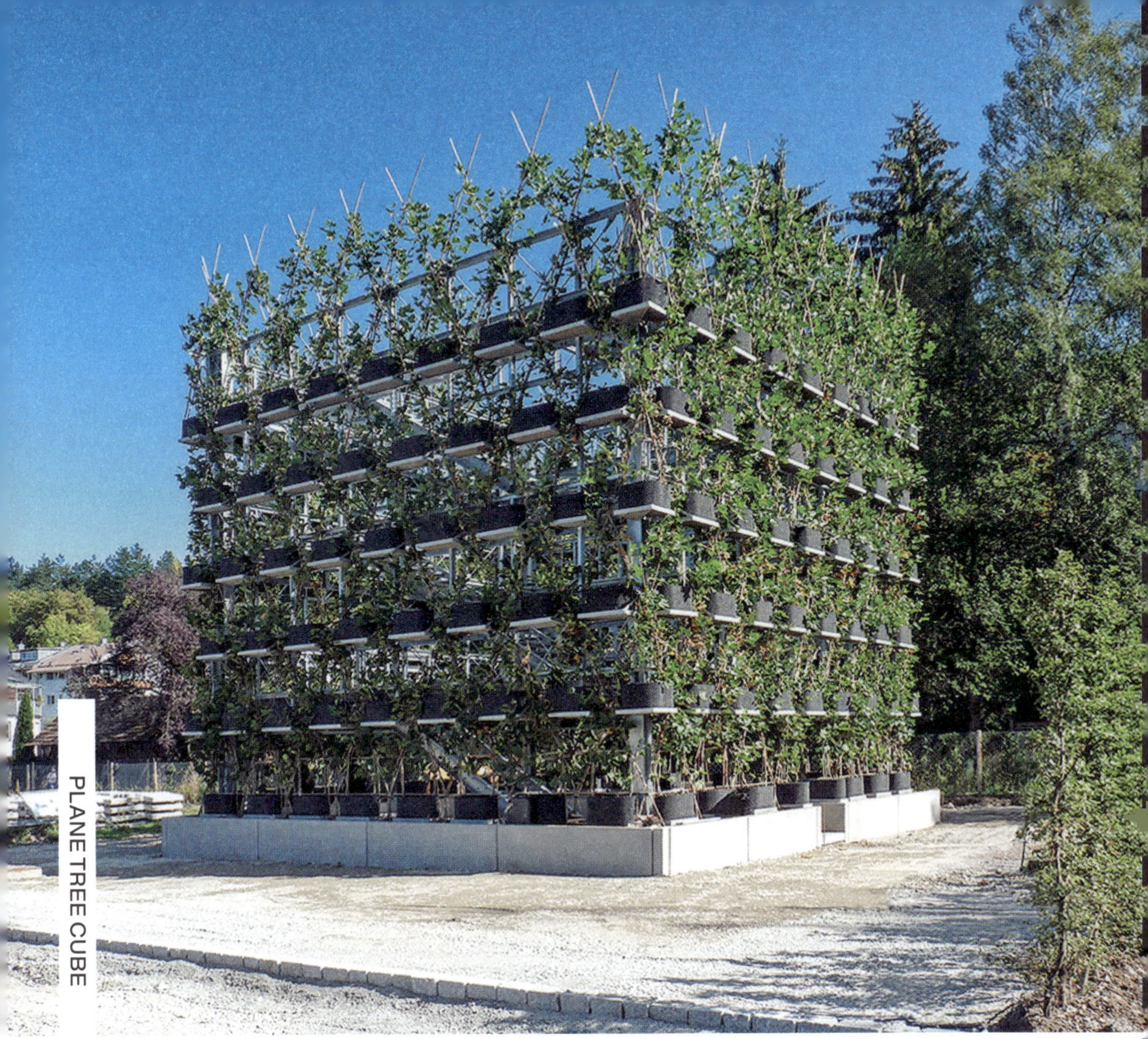

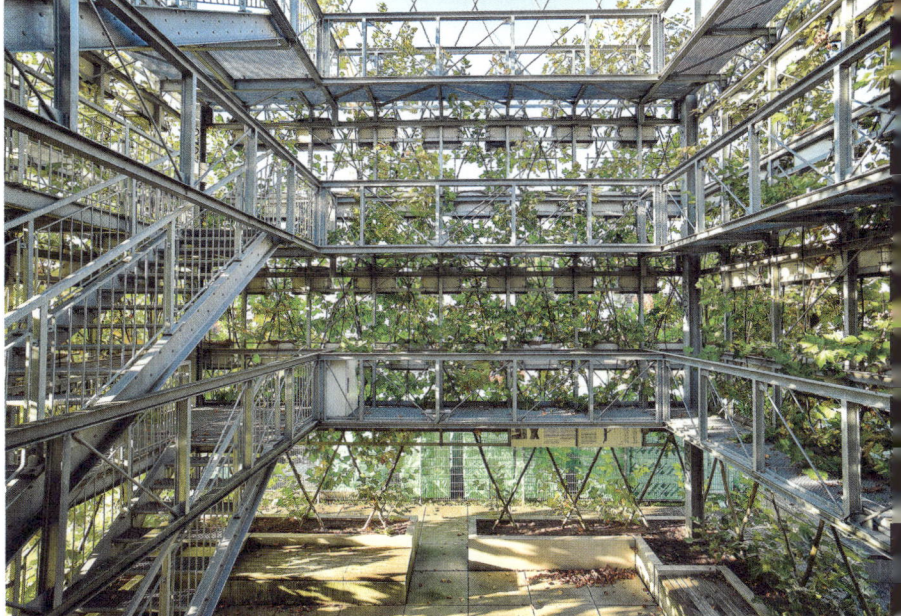

↑ Plane tree cube immediately after structural completion.

↖ Expected development of the structural details: Merging of the trunks, ingrowth of the technical parts and growth in circumference (ca. 15 years).

← Serial cross sections through inosculated branches.

→ Internal view, 2019.

green volume of a fully grown tree even at the beginning. In this initial stage, London plane trees (*Platanus hispanica*) arranged in planters on six levels formed green walls that enclosed a space open to the sky. Inside, service walkways for the gardeners were situated alongside the green walls, and visitor platforms were arranged on three levels.

The trees were joined together through grafting by inosculation in such a way that they fuse into one single organism. In the long term, the living structure can supply itself with water and nutrients from the ground, allowing the planters and irrigation equipment to be deconstructed. The entire structure is supported by vertical steel columns, which are planned to be removed once the living structure has become stable enough to withstand all occurring loads. The visitor platforms are attached to circumferential steel trusses designed to transfer loads to the plant structure at certain points. This takes into account the fact that the Baubotanik structure is expected to develop varying local strengths due to differing degrees of growth in thickness and an anticipated death of trunk sections.

In the first five years, the development of the project was determined by various setbacks and continuous refinement of the technical components, which have to withstand extreme conditions due to public use and the harsh weather at the border of the Black Forest region. Subsequently, the plants developed vitality and showed vigorous growth, trunk diameter gain and sound development of the Baubotanik joints. In the course of further development, the "interior" space will close up more and more as the tree canopy develops, while in the lower area the interwoven trunks will become more prominent, growing thicker and more gnarled over the years to come.

↑   Expected spatial development (ca. 15 years).

↓   Inosculation after several years of growth.

Structural Textile

Team

Natalija Miodragović, Nelli Singer, Daniel Suárez, Prof. Christiane Sauer
Cluster of Excellence *Matters of Activity. Image Space Material* Humboldt-Universität
zu Berlin in collaboration with Weißensee School of Art and Design, Berlin, Saxon Textile
Research Institute, STFI Chemnitz and Max Planck Institute of Colloids and Interfaces,
Department of Biomaterials Potsdam-Golm

Context

Research project as part of Cluster of Excellence *Matters of Activity. Image Space Material*,
exhibited at Humboldt Forum Berlin, Humboldt Lab "After Nature", 2012-ongoing, and at
TAT Berlin, exhibition "Stretching Materialities", 2021–22

Material

Rattan (*Calamus spp.*), willow (*Salix americana spp.*) and other plant-based fiber for
continuous yarn: flax, hemp, wood wool, Kemafil® technology, coarse-warp knitting, wood
fiber hydro-forming on custom-made boards

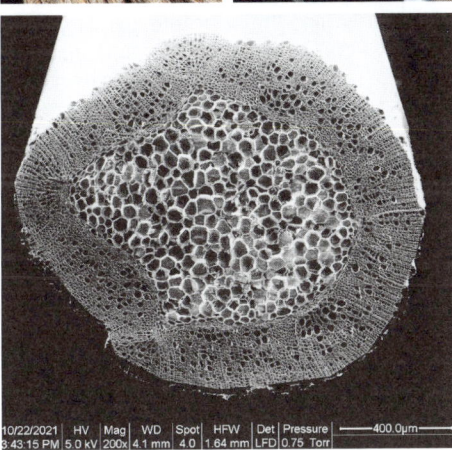

Structural Textile is a transdisciplinary research
collaboration between architects, textile designers, microbiologists, and materials scientists, exploring the upscaling of textile construction techniques to the architectural scale. The
focus is on "active yarns" — filaments that store
elastic energy — such as willow and rattan. The
goal is to preserve — and program — this property when producing continuous yarns as well

↑ The finite branches are
bundled and sheathed
with interlaced yarns in
order to make a con-
tinuous filament with
Kemafil technology. The
willow yarns exhibit a
self-similarity through
overarching scales from
micro to macro.

→ The large looping seg-
ments of the continuous
yarn are obtained with
the coarse warp knitting
machine (STFI).

as large-scale fabrics. The finite willow branches are bundled and sheathed with four interlaced bio-based yarns (e.g. hemp, flax), in order to make a continuous yarn. The bioinspired yarn is "self-similar" — it mimics the microscopic structure of the willow branch.

Willow and rattan plants are readily bendable and formable materials due to the filament structure, allowing them to be knitted. The material thickness determines the bending radius of the knit loop, which can be further reduced through wetting of the fibers. The filaments stay in form when dried, and the material is plasticized to form looping segments. A digital model anticipates formal results and informs the whole fabrication process. On the scale of the fabric, samples of spatial textile are produced from filaments with the coarse-warp knitting technique. Through the use of different filaments or yarns, the fabric behavior such as bounciness, responsiveness, and stability can be programmed.

The project is a contribution to standardization and use of organic non-uniform materials in design and construction, in the context of a post-waste society and circular economy. We envision this knitted structure as a possible fibrous scaffold for composite material in combination with earth or even grown with fungi or bacterial biofilm. The different porosity and densities achievable with knitting allow us to imagine the structure as a filter for light, gaze, sound, and also water and air. The voids in the textile structure are considered as essential for the performance of the whole as its solid elements.

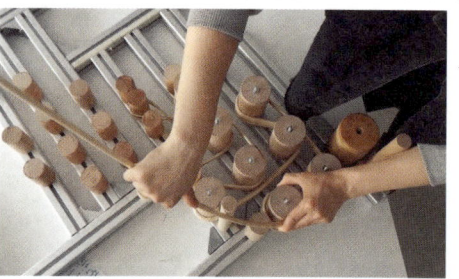

← The wetted material is dried on pre-forming boards with a loop radius related to the material thickness. To research these relations, an adjustable board is developed based on the pre-forming boards by Nelli Singer for the project *Living Beings* (see p. 90).

→ The obtained surfaces can become a structural envelope or scaffold for biomaterial composites.

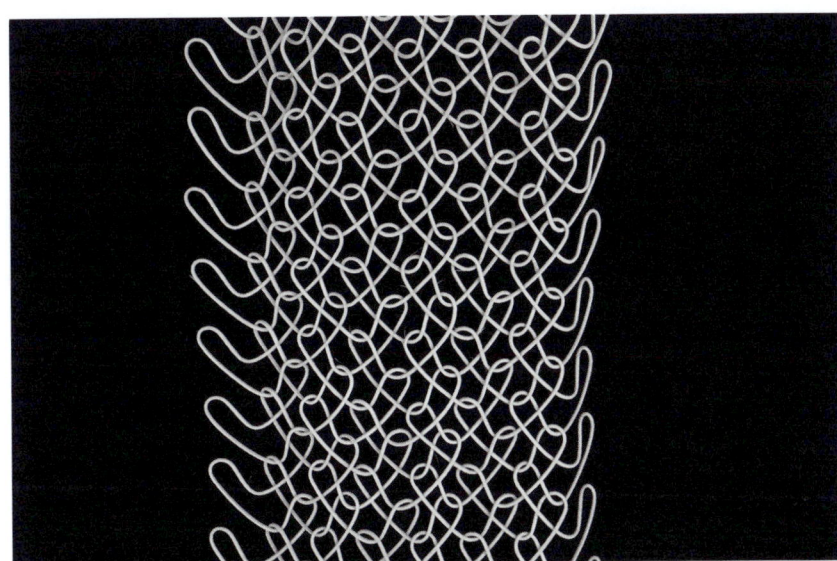

→ Alternating loops up and down, 0 and 1, achieves different structural geometries like convex or concave, tubular or bi-directional on various scales. The knitting density, or resolution, can be used to direct, reinforce or dynamize the surface.

STRUCTURAL TEXTILE

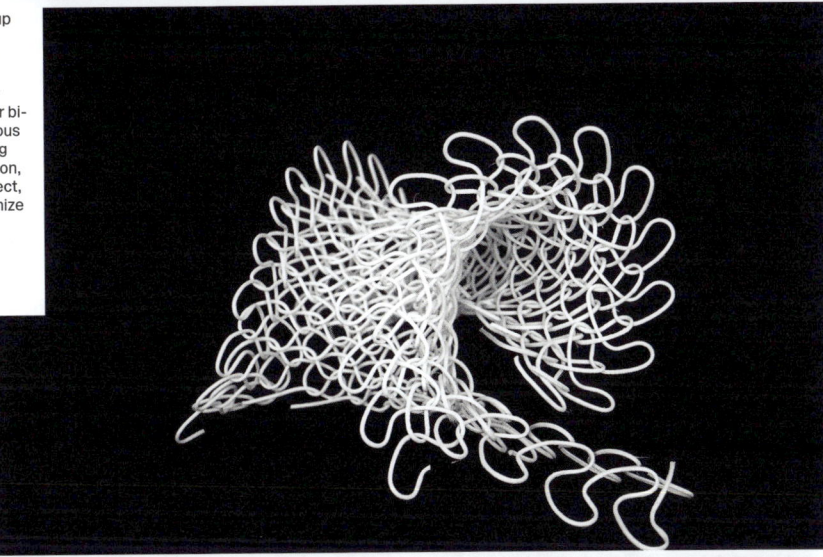

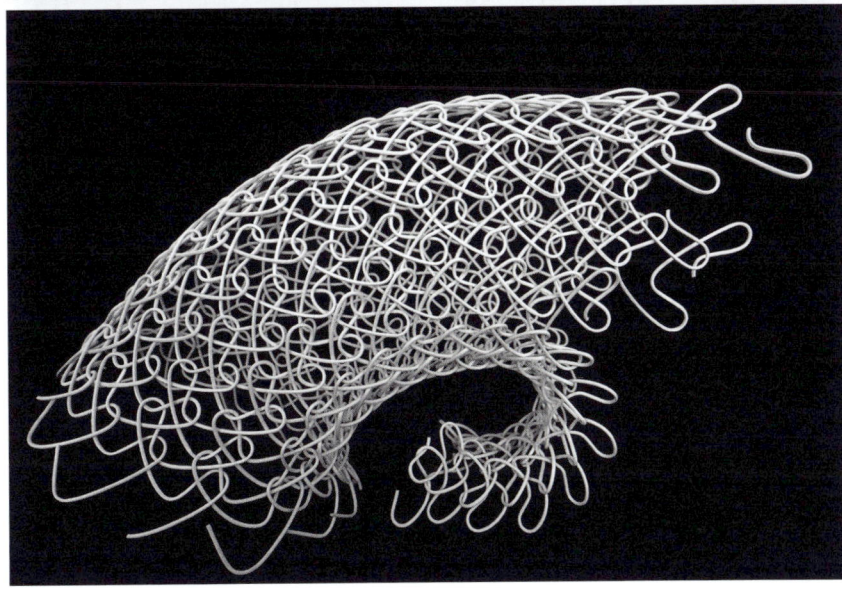

Hybrid Tower

Team

> Prof. Dr. Mette Ramsgaard Thomsen, Prof. Martin Tamke, Yuliya Sinke Baranovskaya, Anders Holden Deleuran, CITA—Centre for Information Technology and Architecture, Royal Danish Academy—Architecture, Design, Conservation, Denmark
> Structural Engineering: Prof. Dr. Christoph Gengnagel, Dr. Riccardo La Magna, Michael Schmeck, University of the Arts Berlin, Germany
> Textile Testing: Prof. Dr. Natalie Stranghoehner, Dr. Joerg Uhlehmann, University Duisburg/ Essen, Germany
> Textile Production: Filipa Monteiro, Jorge Vieira, AFF—A. Ferreira & Filhos/ Vizella, Portugal

Context

> Research project within the Complex Modeling research framework at CITA, exhibited within the Contextile festival in Guimaraes, Portugal, supported through the European COST Action "Novel Structural Skins", 2016

Material

> Knitted Fabric (DyneemaTM), glass-fiber-reinforced polymer (GFRP) rods, fittings (HDPE) Processing by CNC knitting machine (Shima Seiki) interfaced with Rhino/Grasshopper/ Python design environment

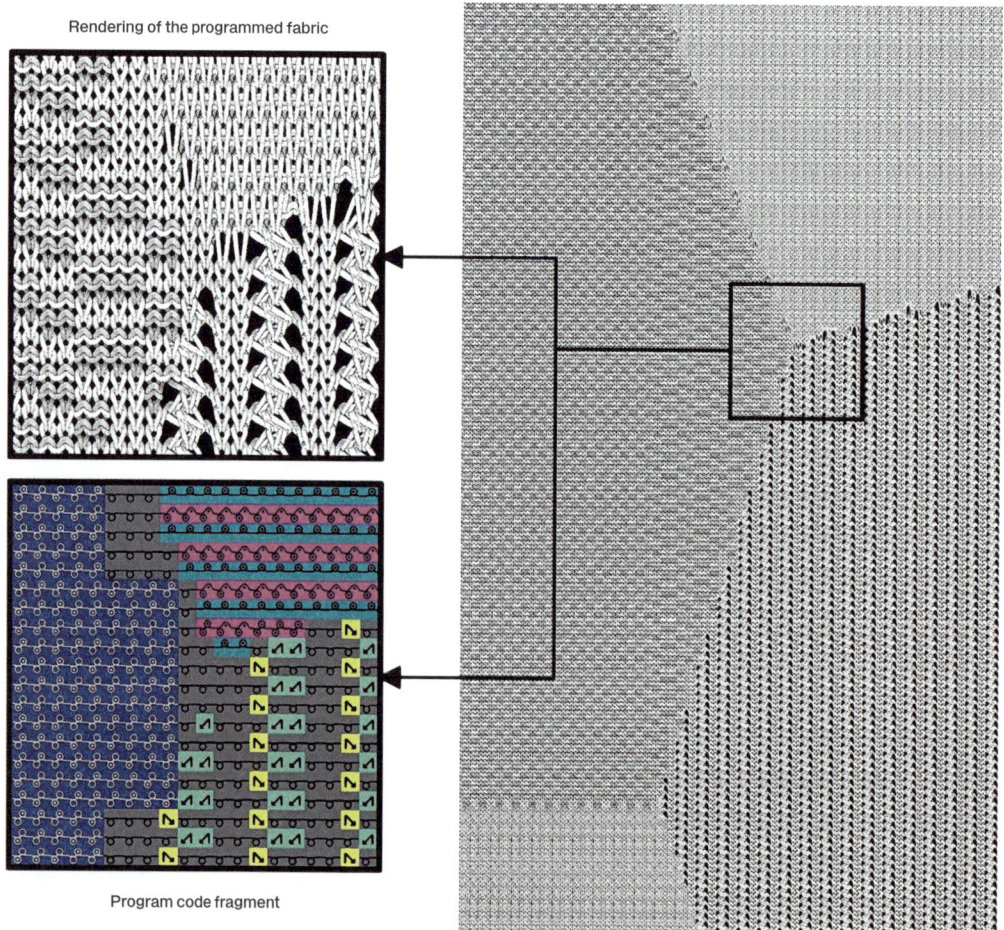

Rendering of the programmed fabric

Program code fragment

Example 1

Traditional thinking in architecture and engineering alike understands the built environment as static. Built structures resist changes in the environment through stiffness. In contrast, *Hybrid Tower* embraces the idea of resilience and adaptation through soft structures. *Hybrid Tower* investigates a very light yet stiff custom-knitted structural membrane in which active bent GFRP rods are embedded, manufactured with bespoke material properties and details. The relationship between skin and structure is a central concern in the field of architectural textiles, positioning the textile membrane either as a cladding skin, or engaged in hybrid dependencies in which membrane and scaffold act as an integrated structural system.

The application of knitted fabrics on an architectural scale requires a prediction of the overall structural performance to guarantee stability. A novel simulation process that interfaces a projection-based relaxation method with a finite-element (FE) simulation extends existing approaches and can effectively simulate the interaction of the constraining fabric, the many pre-tensioning actively bent GFRP rods and external forces. All details are embedded in the material itself and the final shape is directly knitted on the CNC knitting machine. For this purpose, CITA and AFF developed an interface between the design environment and the CNC production of Shima Seiki machines. This enables the direct creation of the machine code that defines the knitting beds, yarn routing and holding patterns, thereby controlling the creation of the knitted textile and providing direct control over structure, material and shape.

initial input mesh · details layer · details clustered by pattern type · pixelated CNC knitting file

← Varying surface properties of knit programming. View of pattern rendering and program code fragment.

↑ Stages of knit pattern design.

↓ Detail of custom-knitted membrane with embedded active bent GFRP rods.

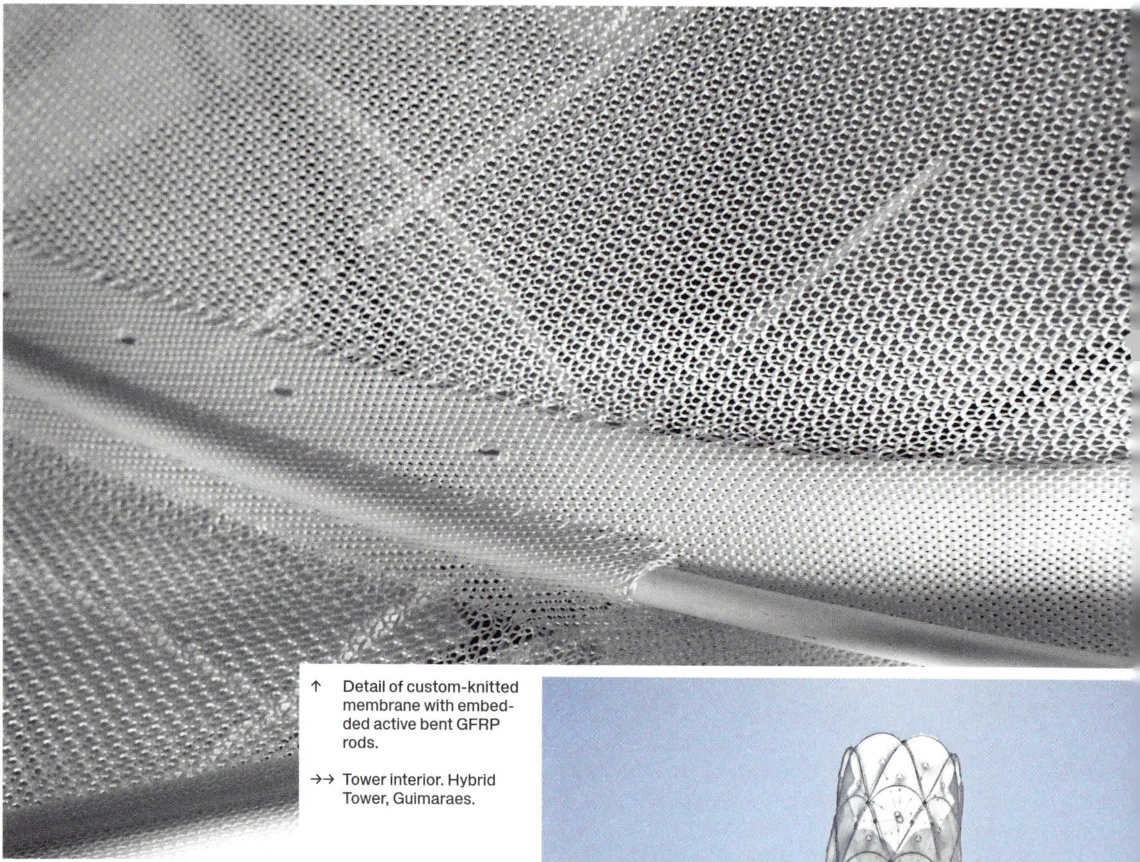

↑ Detail of custom-knitted membrane with embedded active bent GFRP rods.

→→ Tower interior. Hybrid Tower, Guimaraes.

A soft architecture requires new thinking in terms of detailing and assembly. Instead of stacking stories on top of each other, the structural skin of the tower is produced on the ground as a large pre-stressed panel. This is subsequently rolled into shape, tensioned, transported horizontally to the site and erected to the vertical. The nine-meter high structure is so light that it can be carried by only six people. This assembly process is accommodated by a novel type of connectors. These slide towards each other and fix the rods in predefined locations and can withstand vertical loads of up to fifty kilograms each. As a proof of concept, the *Hybrid Tower* was built on the Largo do Toural — the central square of Guimaraes, Portugal. Throughout the three-month exhibition period, the tower demonstrated the strength and high performance of bespoke knit material on a large scale. It provided an exciting architectural intervention for the town, bringing a new experience of a translucent and haptic textile architecture into the urban environment.

The authors wish to thank David Anders Leon, Irina Maximovich, Ida Friis Tinning, CITA, and Raquel Carvalho, Fibernamics.

Maison Fibre

Team

ICD, Institute for Computational Design and Construction, University of Stuttgart:
Prof. Achim Menges, Niccolo Dambrosio, Katja Rinderspacher, Christoph Zechmeister,
Rebeca Duque Estrada, Fabian Kannenberg, Christoph Schlopschnat
ITKE, Institute of Building Structures and Structural Design, University of
Stuttgart: Prof. Dr. Jan Knippers, Nikolas Früh, Marta Gil Pérez, Dr. Riccardo La Magna,
Lab support: Aleksa Arsic, Sergej Klassen, Kai Stiefenhofer
Student Assistance: TzuYing Chen, Vanessa Costalonga Martins, Sacha Cutajar,
Christ van der Hoven, Pei-Yi Huang, Madie Rasanani, Parisa Shafiee, Anand Nirbhaybhai
Shah, Max Benjamin Zorn
In collaboration with FibR GmbH, Stuttgart: Prof. Moritz Dörstelmann, Ondrej Kyjanek,
Philipp Essers, Philipp Gülke with support from Erik Zanetti, Elpiza Kolo, Prateek Bajpai,
Jamiel Abubaker, Konstantinos Doumanis, Julian Fial, Sergio Maggiulli
Supported by the Cluster of Excellence *Integrative Computational Design and Construction
for Architecture* IntCDC, University of Stuttgart

Context

Two-story, robotically fabricated building demonstrator
17th International Architecture Exhibition — La Biennale di Venezia, 2021

Material

Glass- and carbon-fiber composite structure, 23 km glass fibers, 20 km carbon fibers,
robotic winding technology

↑   Fabrication Setup.

→   Exterior view.

↗↗   Full Set of Building
       Components.

MAISON FIBRE

The installation *Maison Fibre* at the Venice Architecture Biennale 2021 presented an alternative approach to conventional construction: the entire structure, including load-bearing floors and walls, consists solely of glass- and carbon-fiber rovings—bundles of endless, unidirectional fibers. The rovings are assembled into fibrous modules using a digitized, robotic winding process—a technology based on a decade of research in fiber construction at the University of Stuttgart.

The fibers are placed very precisely on a winding frame by the robot. Their orientation, alignment, and density can be calibrated to correspond to structural requirements, which enables a radically lighter construction in comparison to massive tectonics. Each building element is made with just a few kilos of construction material. The code-compliant, load-bearing upper-floor fiber structure weighs just 9.9 kilograms per square meter; the wall elements are even lighter.

Maison Fibre takes its inspiration from fibrous structures in biology, such as bone tissues, that are simple in materiality but complex in form. This biomimetic strategy results in an anisotropic structure that minimizes material and is exceptionally lightweight, thus reducing the consumption of energy and material resources during production and transportation.

The system of modular elements can be expanded and adapted to be integrated even within existing architecture. The compact robot production can carry out the entire production locally on site, without producing any significant amount of noise or waste. Developments in material research will make it possible to exchange the glass and carbon for more sustainable renewable natural fiber systems such as flax in the near future.

↑ Detail.

← Assembly at Arsenale, Venice.

MAISON FIBRE

Stone Web

Team

Idalene Rapp, Natascha Unger, Weißensee School of Art and Design Berlin, Department of Textile and Surface Design, Master Studio Prof. Christiane Sauer

Context

Design Prototype, Master Thesis, 2018

Material

Basaltfiber yarn 300 tex — 4800 tex, biobased resin, hand-winding technique

Stone Web is based on the igneous rock basalt, which occurs in large quantities in the earth's crust. In an industrial process, basalt rock can be melted at around 1,500° C, drawn into ultra-thin filaments and further processed into workable fiber bundles, so-called rovings, or into spun yarn. The silicate-based fibers are used as technical reinforcements, e.g. in polymer or concrete composite material. Their mechanical properties lie between the more common glass and carbon fibers. Basalt fibers exhibit further advantages in a high UV-, temperature-, and alkali-resistance.

Stone Web highlights basalt fiber as the main design element and reverses the heavy and solid character of its origin: stone turns into a lightweight material. To produce the filigree modules, basalt yarn is impregnated with bio-based liquid resin and wound onto a removable frame. After curing, this frame is removed and the resulting geometry exhibits a high load-bearing capacity while staying elastic within its lateral net-like surfaces. This results in modules that are both elastic and stable. They are based on the shape of a truncated octahedron and are made of 128 meters of yarn, varying in thickness (from 300 tex to 4,800 tex) and weight (from 45 grams to 950 grams). In load-bearing tests, the strongest modules could carry up to 900 kilograms, which is equivalent to the weight of a small car. The combination of different module types creates spaces that may range from closed to filigree and transparent. The system can take on a variety of shapes and functions. Due to the high tensile strength of the fiber, stable structures can be achieved with minimal use of material. The basalt fibers' weather- and UV-resistance even makes outdoor applications possible.

STONE WEB

135 The pictures show an installation consisting of 100 modules, which are detachably connected and can be easily reconfigured by the user. The weight of the individual components ranges from 45 to 950 grams, so that the entire displayed installation adds up to only forty-five kilograms. The system is extremely light to transport and can be adapted to different spaces and requirements due to its modularity.

The design is reminiscent of growing, cellular structures. Its boundaries oscillate when one moves in and around it. A hybrid typology of space emerges that dissolves the common notions of wall, room, and furnishing in favor of an adaptable structure that inspires new ways of appropriation, use and adaptation. The fiber is more than just material: it creates a new aesthetic and functional approach.

STONE WEB

Concrete Textile

Team

Anne-Kathrin Kühner,
Weißensee School of Art and
Design Berlin, Department
of Textile and Surface
Design, Master Studio
Prof. Christiane Sauer
Cooperation Partners: Saxon
Textile Research Institute,
Chemnitz (STFI), G.tecz
Engineering GmbH, Kassel

Context

Design Prototypes, Master
Thesis, 2016

Material

Composite concrete yarn,
warp-knitted textile, high-
performance concrete
knitted, woven, and knotted
surfaces

The concrete yarn was worked into different prototypes: woven, knitted, and knotted. The final forming of the soft surfaces was carried out by either draping them over support elements or by shaping them in a hanging state through gravity while curing.

An undulating knitted surface was created as an open, mesh-like structure with a strong textile tactility that exhibits stability and load-bearing capacity. As a semi-open permeable concrete wall, it could allow cooling breezes to pass through while providing homogenous shading and privacy — a beneficial combination of properties in hot and humid climates. A weave generates a dense structure that proved to have the highest load-bearing capacity of all samples. Catenary shapes were achieved by hanging the surfaces during

CONCRETE TEXTILE

Even a solid material such as concrete, which stands for stability like no other, can be reconsidered as textile. A "concrete yarn" was developed by Anne-Kathrin Kühner in order to create freestanding and self-supporting textile structures. The composite "yarn" is constructed from a custom-made textile tube that is filled with dry cement and worked into an architecturally scaled surface. After being draped in a dry state, it is watered. When cured, it forms a stable concrete structure that — depending on the type of textile binding — exhibits specific structural and aesthetic properties.

In contrast to common fiber-reinforced concrete, where fibers are mixed into the concrete matrix, here, the outer textile sleeve encloses the inner concrete. The fabric serves as a lost formwork and bonds structurally with the concrete to take on the tensile forces at the strand's outer margin while the concrete core absorbs stress in the center. The composition of the textile, its elasticity and chemical resistance, together with the specific fine-grained recipe of the concrete, were developed through an iterative and experimental design process.

←← Knotted flexible surface.

↖↖ Yarn testing prototypes.

↖ Knitted freestanding screen.

↓ Load-bending test of concrete textile.

hardening. When turned upside down after curing they become structurally efficient forms. By knotting, even a flexible concrete surface could be achieved: the cement is displaced when tightening the knots. At these points the textile remains as a permanently flexible hinge and a tessellated concrete surface is created that can be rolled up and transported easily.

These prototype objects have truly fascinating properties: the concrete is rigid, while its appearance and tactility is soft. Common material attributions are challenged: textile solidifies and becomes structural, whereas concrete is shaped like a textile. Besides hanging and draping, three-dimensional structures can also be achieved directly in the manufacturing process with multilayered textile constructions. Adjusting the yarn diameter allows for

← Three-dimensional knitting.

→ Woven (top), knotted (middle) and knitted (bottom) concrete structures.

↖ Gravity-formed concrete weave.

further scaling and functions, which reveals a range of possible applications.

Since the sleeves remain as lost formwork on the strands, the process does not require any additional molding, which is often a material-consuming and costly factor in concrete construction. The strength of the concrete yarn achieves a stable construction with only small amounts of cement. This approach transforms concrete from a mass material into a valuable design material.

Architectural Yarns

Iva Rešetar, Christiane Sauer

TEMPORAL STRUCTURES FOR
ENVIRONMENTAL REGENERATION

Buildings and their interiors have different temporalities. Beyond the facades and the various constructive layers there exists a multitude of small-scale architectures that make up the living space and which are in a constant state of flux. Furnishings and textiles, for example, are part of these transformations taking place on the inside of a building. Although considered marginal compared to the solid fabric of the building, they go beyond the one-time event of building design and construction, and often fall outside the realm of planned architecture.[1] In fact, the adaptation of the interior begins the moment the building is completed and handed over to its inhabitants. They participate in an "entropic" activity in which partial adjustments, changes of purpose, appropriation, and reinvention gradually transform the space from the inside out.

Textiles have always contributed to these processes, since they belong to the "stuff"[2] that forms the living environment and mediates its material, sensorial, and symbolic dimensions. In the *Lexikon des künstlerischen Materials: Werkstoffe der modernen Kunst von Abfall bis Zinn*,[3] the authors state that "as well as bodies, spaces were also temporarily transformed by textiles," and refer to the typologies of textile structures such as tapestries or carpets. As elaborate, large-scale objects, tapestries were early examples of fabrics narrating and constructing (material) histories, and as such combined the initial, sheltering function of textiles with a tactile and communicative role, unfolding in dialog with the interior space. Modern and contemporary architectural textiles have challenged historical techniques and archetypes, especially their decorative, technological or gendered aspects, defining an open and critical field of work between sculpture, architecture, and textile art.[4]

Architectural Yarns engage in this ongoing negotiation between the scales of textiles and architecture, and between the different timescales and life cycles of building elements. As a design intervention into the existing built space, in contrast to solid construction, they are less concerned with permanence and more with adaptation and flexibility of use.

This kind of intervention is more than a pragmatic response to the critical role of the building industry in the current environmental and climate crises. Faced with the abundance of the existing built and anthropogenic mass,[5] architectural practice may not necessarily be oriented towards new

---

1   Tim Ingold describes how human or non-human activity perpetually shapes the built environment through dwelling: "Building, then, is a process that is continually going on, for as long as people dwell in an environment. It does not begin here, with a pre-formed plan, and end there, with a finished artefact. The 'final form' is but a fleeting moment in the life of any feature, when it is matched to a human purpose, likewise cut out from the flow of intentional activity." T. Ingold, *The Perception of the Environment: Essays on Livelihood, Dwelling and Skill* (London and New York: Routledge, Taylor & Francis Group, 2011).

2   In their observation about the aging and transformation of building components, Steward Brand and Frank Duffy introduce the notion of "Shearing Layers of Change" to reflect on the longevity of everything associated with buildings—from their site and structure to the "stuff"—such as "flighty interiors" and all the "mobilia" with the highest rate of change. S. Brand, *How Buildings Learn: What Happens after They're Built* (New York: Penguin Books, 1995).

3   M. Wagner, D. Rübel, and S. Hackenschmidt, ed., *Lexikon des künstlerischen Materials: Werkstoffe der modernen Kunst von Abfall bis Zinn*, Beck'sche Reihe 1497 (Munich: C. H. Beck, 2002).

4   J. Jefferies and D. Wood Conroy, "Shaping Space: Textiles and Architecture—An Introduction," *TEXTILE* 4, no. 3 (November 2006): 233–37, https://doi.org/10.2752/147597506778691431.

5   E. Elhacham et al., "Global Human-Made Mass Exceeds All Living Biomass," *Nature* 588, no. 7838 (December 17, 2020): 442–44, https://doi.org/10.1038/s41586-020-3010-5.

buildings, instead acknowledging the refurbishment and retrofitting of existing spaces as one of the most important current and future challenges.[6] Not only might the technical standards for existing buildings become outdated, but the requirements of use may also alter. Over recent decades, in densely populated urban areas, more and more office and industrial buildings have been converted into housing. On the other hand, homes often function as temporary offices or places of learning — especially during the current pandemic. Today and in the future, spaces are expected to adapt more rapidly to changing circumstances, as the link between space and program dissolves with increased mobility and flexibility between private and professional life.

With the project *Architectural Yarns*, we explore alternative design strategies for retrofitting building interiors. The term "retrofit" does not usually refer to the design context. It is an adaptation process that involves the replacement and renewal of outdated equipment and, more recently, various measures for improving the environmental performance of buildings, for instance, poorly insulated stock. This process, however, can outgrow its normative and technical character. As a spatial and aesthetic practice, it can be understood as a form of environmental regeneration and repair in which the inhabitants actively participate in modulating the indoor climate and adapting their immediate surroundings. Our approach views soft structures as "clothing" and "enveloping" the interior in a temporary way. Rather than being fixed or concealed within the layers of the building structure, as is often the case with common technically driven retrofit measures, *Architectural Yarns* appear as tangible and adaptable elements of the interior space. In the form of screens, curtains or even freestanding walls, they take on a variety of tasks in shaping and tempering environments, overcoming the usual distinction between the technical equipment and interior furnishings.

What are *Architectural Yarns* and how are they made? As an open design framework, encompassing a wide range of experiments, yarns combine versatile materials into continuous strands on a scale beyond that of conventional textiles. Their production is made possible by *Kemafil®* technology (core-mantle-filament technology) used in the field of technical textiles for sheathing loose materials — mainly for the production of rope- and strand-like structures for geotextiles. *Kemafil®* was patented in the 1970s in the former German Democratic Republic by the Saxon Textile Research Institute, STFI Chemnitz (see p. 99). The idea is to bind textile residues together to form a new and strong rope system — an early concept of upcycling textile waste. With a diameter in the centimeter range, these tubular elements can be filled with loose granulate (such as earth) or fiber material (like flax). In our research project, this technique is transferred into an architectural context by introducing yarn as a flexible and reconfigurable building element — either structural and self-supporting, or with specific functions, such as climate or acoustic modulation. Some of these experiments are described in the following design studies involving thermally active phase-change materials (Figs. 1-3) and flax fibers (Figs. 4-8), as well as in the series of examples including earth, textile waste, or wool as bulk materials (see *Scaling Fiber* Case Study p. 149).

6   Architect Anne Lacaton sees reuse and transformation as a call to redefine the way the profession deals with climate and ecological urgencies: "Transformation is the opportunity of doing more and better with what is already existing. The demolishing is a decision of easiness and short term. It is a waste of many things — a waste of energy, a waste of material, and a waste of history. Moreover, it has a very negative social impact. For us, it is an act of violence." O. Wainwright, "'Sometimes the Answer Is to Do Nothing': Unflashy French Duo Take Architecture's Top Prize," *The Guardian*, March 16, 2021, sec. Art and design, https://www.theguardian.com/artanddesign/2021/mar/16/lacaton-vassal-unflashy-french-architectures-pritzker-prize.

Fig. 1
Thermal yarn filled with granular PCM material.

Fig. 2–3
Thermal Screens with integrated PCM yarns, processed
by *Kemafil®* and coarse
warp-knitting technologies,
creating a flexible thermal
boundary.

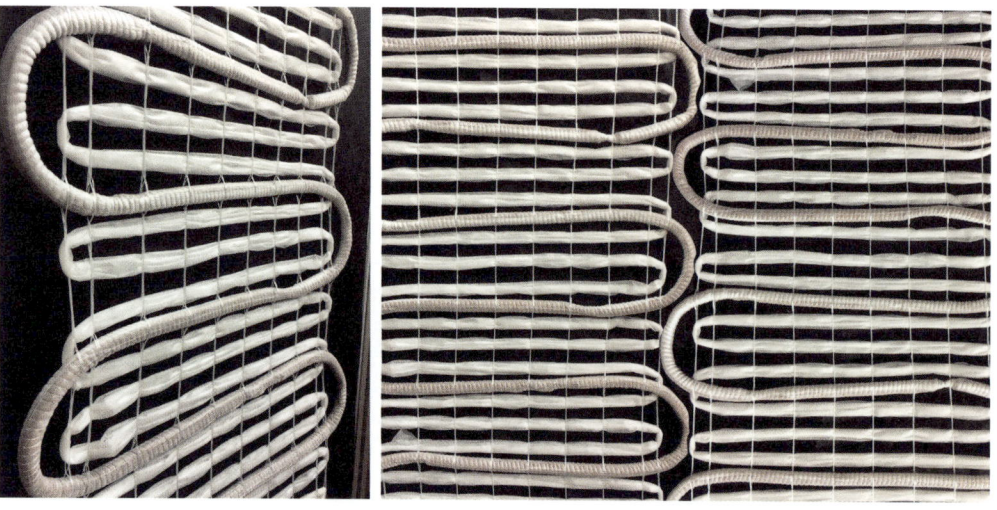

This study investigates the design potential of bulk yarns for indoor temperature modulation. As granular infill, we used the smart, thermally active *Phase Change Material* (PCM) to create lightweight and fibrous "environmental devices" (Fig. 1). PCMs have an ability to affect the ambient room temperature by changing states between solid and liquid in repeatable melting and solidification cycles. When the room temperature increases, they melt and draw thermal energy from the surrounding air, thus cooling their environment; when the temperature drops below a defined point, they solidify again and release this thermal energy — thereby warming their environment.

PCMs are already available, integrated into building components, for instance in gypsum panels, glass panels or masonry blocks. Such elements provide a substantial planar surface for thermal exchange. In comparison, our textiles constructed from tubular yarns maximize the exposure of the PCM to its surrounding owing to their geometry with a higher surface-to-volume ratio. Therefore, better performance can be achieved with less material. Following the guidelines on the effective thickness of PCM, which supports full melting and solidification,[7] yarn diameter was dimensioned to twenty millimeters, and spacings in between were introduced to facilitate good air circulation and therefore thermal exchange. Compared to the PCM coatings applied in the production of existing functional textiles, the upscaled yarns enable a significant increase in thermally effective mass.

The resulting structures can be used as easy-to-handle sliding screens that can be positioned according to climatic requirements, creating a flexible thermal boundary (Figs. 2-3). Since window areas and large glass surfaces in particular provide additional heat input in buildings, interior screens offer a suitable temperature modulation in addition to external sun protection. In other scenarios they can also become moveable room dividers or wall coverings.

The adaptiveness of the interventions is reflected in the logic of material assembly and disassembly embedded in the manufacturing process: the structures are made by filling sleeves bound by a looped thread with granular material, allowing them to be separated easily into their initial components at the end of their life cycle. Entire elements can be moved from one place to another. This is a major advantage over conventional built-in PCM components, where the materials are bonded together and permanently integrated into the building structures. With textile screens, thermal activity is no longer hidden "behind the wall," but instead becomes a visible element of the interior, inviting engagement and participation.

TANGLED WALL WITH FLAX FIBERS

The second case study demonstrates the possibility of building differently — with plant fibers. In this experiment, various textile-manufacturing techniques, both industrial and manual, are explored to create soft, provisional architectures — room partitions, enclosures or interior landscapes. Flax fiber and cotton cord are processed into continuous yarns, flexible enough to be entangled by hand in a do-it-yourself manner. Inspired by traditional techniques of knotting and crocheting, these entanglements resemble the making and unmaking of ordinary clothing, where the structure can be temporarily given a shape, only to be unraveled later.

ARCHITECTURAL YARNS

7   VDI, "VDI 2164, PCM Energy Storage Systems in Building Services" (Düsseldorf: Verein Deutscher Ingenieure, December 2016).

Environmental regeneration plays a major role in the choice of flax—a coarse, inexpensive industrial plant and an unconventional building material. Flax fibers are obtained from the stem or bast of the plant and, together with hemp, are considered the least ecologically harmful of all cultivated natural fibers.[8] Flax is undemanding: grown without fertilizers and pesticides, it is a plant that does not require high-quality soils and it does not compete with agricultural land. To release the fibers, the plant stem is treated by retting or soaking in water, a process that allows the woody parts to decompose slowly. Both flax and hemp fibers have recently been investigated as a promising alternative to cement and gypsum products, which are ubiquitous in interior construction. However, these examples are often based on fibers pressed into solid boards with a homogenous structure, leaving the novel potential for the design with the plant matter untapped and open to alternative, textile-oriented approaches.

Assembling the *Tangled Wall* is a tactile practice involving crossing, intertwining, and knotting yarns to shape the formless bulk into a textile structure, with many possible arrangements. To guide this hands-on process, flax yarns are produced in segments of varying thickness, either loosely or tightly wrapped with cotton strings.

8   K. Slater, *Environmental Impact of Textiles: Production, Processes and Protection* (Cambridge: Woodhead, 2003).

Fig. 4
Flax yarn is wrapped with cotton threads in a pattern providing the "manual" for wall making.

Figs. 5–6
The making of a *Tangled Wall*.

ARCHITECTURAL YARNS

Fig. 7–8
Architectural yarns with
loose fibers are worked into
a continuous yarn by *Kemafil®*
sheathing.

Here, the "manual" for making the wall is embedded in the structure of the yarn: the transitions from thin to thick serve as annotations for scaling and interlacing loops, and then appear as patterns in the wall fabric, revealing the process of making (Figs. 4-8).

Such bodily engagement with the textiles conveys a "tangible thinking,"[9] similar to how Jessica Güsken pictures handcraft techniques of tapestries and carpets, particularly the concept and practice of knotting, as examples of structures "overflowing and potentially always permitting new, further stitches and connections of the most diverse kind."[10] But this technique, she notes, does not merely provide examples. Rather, it serves as a model of "mixed layers and intermingled bodies,"[11] in the case of our experiments, a sensual mode of interconnecting plants, things, bodies, and environments, which find their contact surface in a fiber, yarn, and the built fabric.

9   J. Güsken, "Knoten: Lösen, Knüpfen, mit der Haut denken. Michel Serres' tangible Philosophie der Gemenge und Gemische," in *Michel Serres. Das vielfältige Denken. Oder: Das Vielfältige denken*, ed. R. Clausjürgens and K. Röttgers (Brill | Fink, 2020), 37–57, https://doi.org/10.30965/9783846765142_004.
10   Ibid.
11   Ibid.

ACKNOWLEDGEMENTS:

The research on *Thermal Screens* received support from the Fritz und Trude Fortmann-Stiftung für Baukultur und Material. *Tangled Wall* is a project developed in collaboration between the Weißensee Kunsthochschule Berlin, the STFI Chemnitz, and the Cluster of Excellence *Matters of Activity. Image Space Material.*

Architectural Yarns is a research project at the Cluster of Excellence *Matters of Activity. Image Space Material*, research team: Iva Rešetar, Christiane Sauer, Maxie Schneider, Josephine Shone.

Authors of the text acknowledge the support of the Cluster of Excellence *Matters of Activity. Image Space Material* funded by the Deutsche Forschungsgemeinschaft (DFG, German Research Foundation) under Germany's Excellence Strategy—EXC 2025—390648296.

Scaling Fiber—Experimental Yarns

Team

Saskia Buch, Martha Panzer, Jasmin Sermonet, Clara Santos Thomas, Weißensee School of Art and Design Berlin, Department of Textile and Surface Design
In collaboration with Cluster of Excellence *Matters of Activity. Image Space Material*, Saxon Textile Research Institute (STFI)
Mentoring team: Prof. Christiane Sauer, Ebba Fransén Waldhör, Maxie Schneider, Iva Rešetar, Dr. Bastian Beyer, Dr. Michaela Eder, Dr. Heike Illing-Günther

Context

Design Prototypes, MoA Design Research Studio, *Scaling Fiber—Experimental Yarns*, 2021

Material

Yarns from earth, shredded textile, wool, bound with *Kemafil®* technology

↑   *Wool Yarn.*

The MoA Design Research Studio *Scaling Fiber* explored the idea of *Architectural Yarns* situated between design experimentation and structural development. The transdisciplinary exchange with architecture, cultural science, material science, and textile technology aimed at concepts for sustainable construction. Yarns were scaled into the spatial context as building elements. Their specific composition rendered new joining possibilities like felting, interlocking, and growth. Three material variants—wool, textile waste, and earth—redefine textiles as a structural, social, and spatial framework for activity.

WOOL YARN

Wool has been used for centuries for clothing, but also as a material for insulating buildings. It is little known, however, that sheep's wool is today largely burned as waste. A steadily growing competition makes the trade of (organic) wool for small sheep farms almost

↑   Textile Yarn.

↑|→ Earth Yarn.

↑   Wool Yarn.

impossible. But where to put the surplus wool, which combines so many useful properties? *Wool Yarn* is based on the idea of producing an upscaled, ecological yarn from sheep's wool. A solid three-dimensional bond between yarns can be achieved through mere felting of the woolen fiber without any additional connecting elements. Two different applications were explored: an oversized three-dimensional strand with a cuddly feel creating flexible furnishing that encourages touch and interaction, and an open mesh-like structure hanging in space for a positive effect on the room climate, moisture, and acoustic absorption.

TEXTILE YARN

The idea was to create modular, textile furniture or architectural structures that can be repeatedly assembled in new and different ways by simply clicking the connecting pieces open and closed. The textile materials create soft, flexible spatial elements. The yarn consists of textile waste regained through an industrial recycling process, which is soft and loose but obtains stability by being sheathed and laced with elastic cords. Because the continuous elastic cord also serves as a connection device, the yarn can be joined together at any point throughout its length to achieve different configurations. The spatial elements can be used in different ways: flat, as a carpet or mattress or three-dimensional, as a bench, sofa or seating group.

EARTH YARN

As a speculative material, *Earth Yarn* addresses questions about the future of habitats as mediations between humans and the landscape. In this experiment, low-cost materials such as earth, seeds, cellulosic sheathing, and threads are used. Different types of soil can be adapted to different needs and plants. The growth of the seeds incorporated into yarns creates habitats that could be described as small, species-rich biotopes. This idea becomes increasingly important as the number of insects and, especially bees, declines. The *Earth Yarn* is permeated by water, light, and warmth, and follows a natural life cycle. It creates biotopes where plant growth would otherwise be difficult. Vertical gardens can be constructed on unused urban walls, and can provide niche spaces for insects. The yarn might also be used indoors to improve the climate.

Fabricating Space

Creating space with fabric is basic to human culture. To this day, Nomad cultures make use of textile's light weight for transport and of its flexibility for mounting. Simple tents are created from woven cloth on studs, and deployable yurts from thick felt mats on wooden mesh encompassing even large room sizes.

The following case studies show different strategies for creating space from fabric. Folding, looping or tucking facilitate a self-supporting structure without any additional bracing: a weave pattern that is programmed for origami folding, centimeter-thick industrial felt, or spacer fabrics that are worked into architectural structures. Textile draping can also be deployed to form large-scale elements. As a pliable formwork it creates soft geometries that can become permanent when filled with concrete, or temporary when covered with ice. As a curtain, fabric can even cover the height of a facade and become a flexible building skin that adapts to different scenarios of use.

Fabric creates not only architectural space, but also symbolic or social space. It can become a carrier of information, imagination or memory. A communal art project fabricates a network of social and physical spaces through the creation of a quilt. It depicts the history of an urban community and city district through personal memories. A theoretical reflection on framed textiles — swatches — complements this chapter by highlighting them as symbolic material that can expand through their physical presence into the world of the virtual, the "elsewhere."

ArchiFolds

Team
> Studio Samira Boon, Amsterdam / Tokyo, commissioned by and in collaboration with TextielLab, Tilburg, Netherlands and Prof. Tomohiro Tachi, University of Tokyo, Japan

Context
> Research project later implemented by various clients: Jiaxing Gallery, Theaters Tilburg, and MORE Architecture, among others, 2014–ongoing

Material
> Paper, polyester monofil, mohair and abaca, woven on a Jacquard loom

ARCHIFOLDS

The textile architecture Studio Samira Boon researches how to transfer the principles of origami—the art of folding three-dimensional paper shapes—to woven fabric. The studio uses the possibilities of complex pattern-making on computer-controlled Jacquard looms to create fabrics that fold along predefined lines. Unlike other textile-pleating techniques, where folds are heat-set or stitched in place, these folds are intrinsic to the structure of the fabric. Through the material combination of different yarns and their arrangement in relation to each other, a tension is created between hard and soft areas in the weave that makes the fabric expand into three-dimensionality once it is taken off the loom.

The fabrics are the result of an extensive research process into paper-folding structures, textile materials, weave bindings and software. To develop suitable folding principles, the studio studied Japanese origami techniques and collaborated with computational origami specialist Prof. Tomohiro Tachi from the University of Tokyo. The transfer to textile involved a close collaboration with technical weavers at the TextielLab, a textile prototyping workshop at the TextielMuseum in Tilburg, Netherlands. The difficulty was to produce a fabric with hard surfaces combined with soft and flexible folding lines—a challenge even on advanced weaving machines. Yarns of varying stiffness and flexibility such as paper, polyester, mohair, and abaca were experimented with to achieve the right combination of structural stability and foldability.

Based on the initial research into three-dimensional fabric structures, the studio has produced a number of lightweight origami fabrics commissioned for specific buildings and interiors. Functioning as flexible architectural interventions, the sculptural fabrics divide space, alter acoustic properties, and regulate light. Because of their structural flexibility, they offer a multitude of installation options. The very compact folding also makes them easy to move and transport.

ARCHIFOLDS

Merging Loops

Team

Bára Finnsdóttir, Weißensee School of Art and Design Berlin, Department of Textile and Surface Design, Design Studio Prof. Christiane Sauer

Context

Design Prototype, 2016

Material

3 mm furniture felt, 10 mm industrial felt, veneer

Merging Loops is a study for creating coherent surfaces from reconfigurable flexible modules by simply tucking them into each other. The modules are cut from flat material and gain stability through the loops that are interlocked back into the surface. The system is applicable at various sizes. Bára Finnsdóttir transferred the principle into a small-scale hanging and a large-scale freestanding version.

The hanging version consists of fork-shaped felt modules that are laminated with thin veneer to enhance the friction of the surface materials when connecting the elements. Depending on the joining pattern, the modules create a distinct rhythm of color and texture, of material and void. The variety of configurations is reminiscent of a coding system embedded in textile like the ancient Khipu strings that were knotted in a specific pattern to transmit information. By reconfiguring the modules, the user can adapt the layout in density, color, and size according to current needs and imagination.

The freestanding design is based on the same plug-in principle. For use as a partitioning screen, not only the module sizes but also the material thickness had to be scaled. A 10 mm industrial felt made from recycled polyester and polypropylene waste fibers was used for this purpose. The interleaved form of the modules is designed to enable production from the roll without any offcuts in order not to waste any of the raw material. The resulting elements are joined by hand through bending and looping the bands to form a free-standing textile wall that can be extended to any length required. If not needed, it can be quickly taken apart and stored, packed or easily transported in a flat state.

The set-up's three-dimensional shape and the fibrous surface result in very good sound-absorbing properties, which makes the structure ideal for use in office floors or conference rooms. This application for a recycled industrial material provides a new aesthetic and functional value and therefore upcycles it through design into an appealing and smart system.

MERGING LOOPS

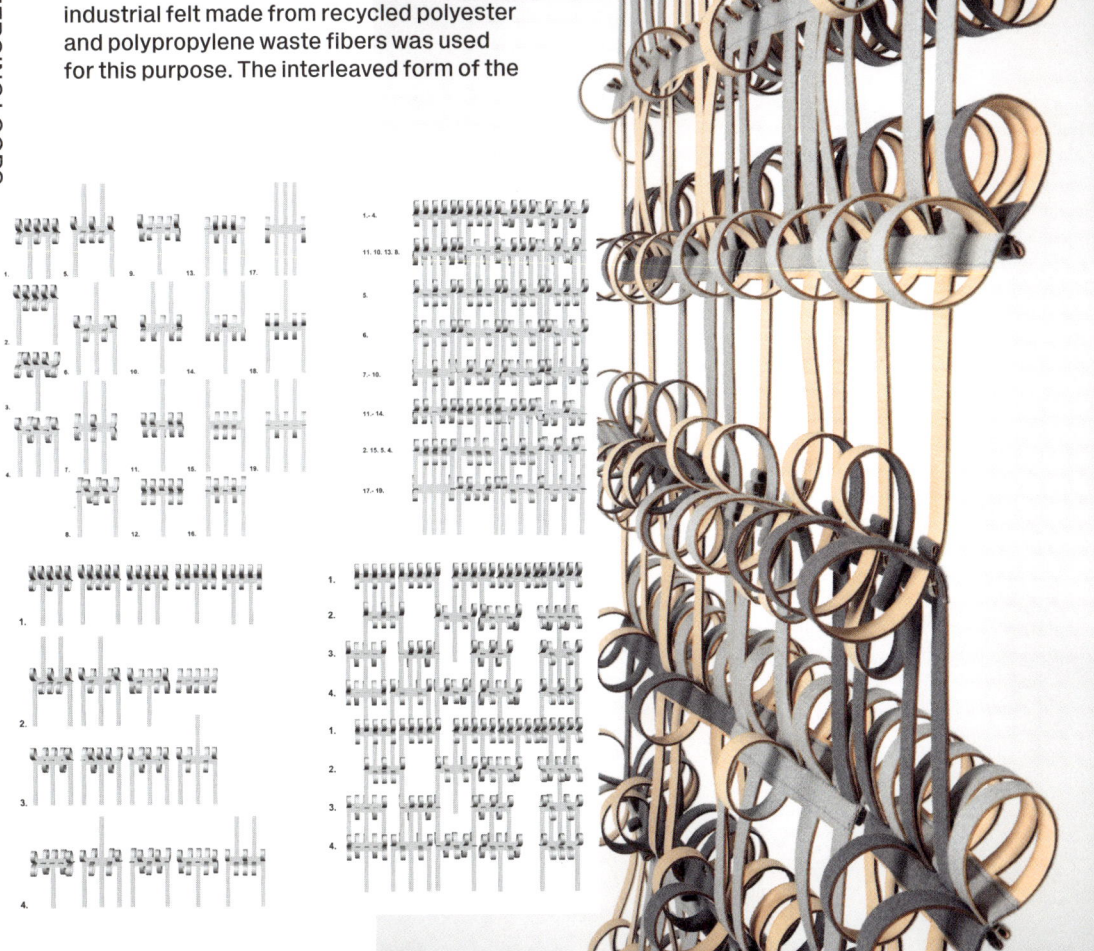

SpacerFabric_Pavilion

Team

Prof. Claudia Lüling, Frankfurt University of Applied Sciences, Frankfurt Research Institute for Architecture • Civil Engineering • Geomatics (FFin)
Student Team: L. Aust, J. Beuscher, S. Biehl, M. Cicala, I. Cursio, J. Dittmann, K. Gregurevic, M. Haas, E. Krücke, N. Lüer, N. Micheev, I. Micorek, M. Simlesa, C. Sotgia, M. Vogel, A. Zgodzinski
FabricFoam Research Team: Steffen Reiter, Iva Richter, Ilga Richter, Natalija Miodragovic, Johanna Beuscher, Prof. Claudia Lüling

Context

Research prototype as outcome of a seminar on spacer fabrics, 2015

Material

Warp-knitted spacer fabrics and spray foam

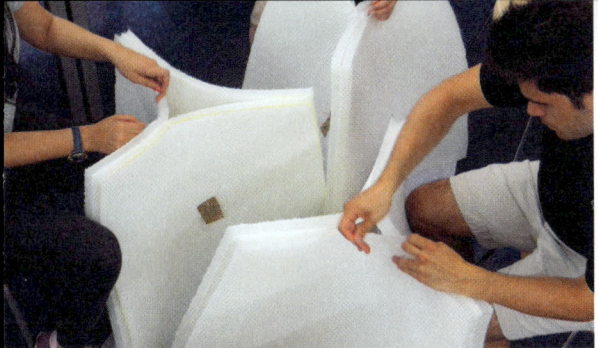

The *SpacerFabric_Pavilion* is a self-supporting dome made of soft, multilayer spacer textiles. Structural stability is achieved through a combination of folding and foaming, with the load-bearing structure forming a unified whole with the membrane. As part of a seminar on spacer textiles at Frankfurt University of Applied Sciences, a group of students investigated how the partial foaming of defined areas in a warp-knitted spacer textile can produce a building skin that is lightweight yet stable in compression and tension, insulating, sound-absorbing, and translucent.

The research included numerous analogous tests on the handling of foam and textiles, in order to explore the material properties and possible workflows for the construction process. Model studies were made on the spatial quality of the pavilion in terms of proportion and comfort. Three typologies emerged from different design concepts: folded, curved, and modularly assembled construction methods. These were further investigated in smaller teams before the group decided on the realization of a larger modular structure.

The final result combines individual, folded pyramid-shaped textile modules, with foamed connections in between, forming a spatial supporting structure. The experimental pavilion shows the design potential of the new composite material combining textile-fibrous parts with foamed-porous components. The resulting dome has a diameter of five meters and a height of three meters. It is airy, translucent and sound-absorbing — inviting visitors to linger.

SPACERFABRIC_PAVILION

Centres for Traditional Music

Team
OFFICE Kersten Geers David Van
Severen, Brussels, Belgium

Context
Public building, renovation and addition,
Muharraq, Bahrain, 2012–18

Material
Stainless-steel ring mesh with motorized
opening system, aluminum folding
doors, concrete building structure

Separating the interior from the exterior, curtains are flexible boundaries that regulate visibility. In the theater, the stage curtain also has a symbolic function; its opening and closing marks the beginning and end of a performance. The *Centres for Traditional Music* in Bahrain by architecture firm OFFICE Kersten Geers David Van Severen, was conceived and built as a space for the cultural traditions of the local community of pearl fishers. It is covered in a seamless-steel mesh fabric that opens like a stage curtain when the building is in use — exposing the communal space on the ground floor where musical performances take place. Marking the threshold between the street and the semi-public community center, the fabric facade simultaneously frames passers-by and invites them in.

The metal mesh also gives shade from the harsh desert sun, completely covering the building from the top. The simple structure of concrete columns, platforms, and glazed walls with additional perforated wooden shutters that can be folded open, provides intimate interior spaces behind the veil. The fine permeability of the ring mesh allows for an unobstructed view to the outside, while letting air pass through.

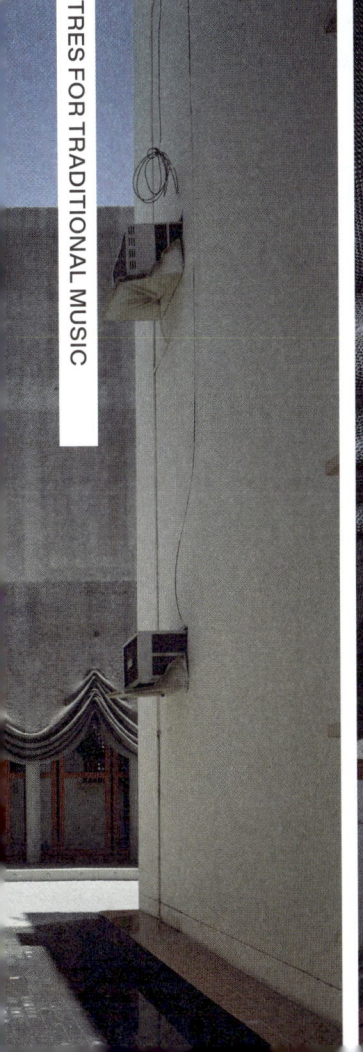

Sagrada Familia in Ice

Team

Ass. Prof. Arno D.C. Pronk, Faculty of the Built Environment, Innovative Structural Design (ISD), Eindhoven University of Technology
Student team: Teun Verberne, Jordy Kern, Faculty Architecture, Building & Planning, Eindhoven University of Technology

Context

Research building by Structural Ice Research Group, Eindhoven University of Technology, Juuka, Finland, 2015
Funded by Tulikivi Oyj, Pekkaniska Oy, and Hitachi Ltd.

Material

PVC-coated polyester fabric, ropes, Pykrete (cellulose fiber-ice composite), air and water pumps, shredders, containers with mixers, hoses, nozzles, crane, inflatable molding

Ice is a well-known material, but as a building material, it is less common. After Arno Pronk and his team carried out research on the material properties and form-finding of reinforced ice, the *Sagrada Familia in Ice* was realized in January/February 2015 in Juuka, Finland.

The first scientific studies for the application of ice composites were conducted during World War II by Geoffrey Pyke and his team. Since then, research has been carried out on the application of fiber-reinforced ice, better known as Pykrete. Here, sawdust is mixed with water and sprayed onto the textile membrane with a centrifugal pump and adjustable nozzle. Different studies demonstrated that the addition of (natural) fibers to ice results in a three times higher strength and high ductility compared to regular ice. Pykrete also improves the resistance against thermo-shock that might occur during the building process as a result of spraying water on a frozen shell structure. The pavilions made by Heinz Isler with frozen textiles and ropes show the potential of using textile as a mold in ice structures. Isler also demonstrated that ropes and textile fabrics of natural fibers are very suitable for the form-finding and reinforcement of shells and grid shells.

The design of the *Sagrada Familia in Ice* is based on the Cathedral of the same name in Barcelona by Antoni Gaudí, which was designed using a model of suspended chains, resulting in catenary shapes. A suspended chain or rope will always take the shape of a smooth curve, meaning that the chain is subjected only to tension and no pressure at all. Once the curve of the chain is turned upside down and the whole setting is reversed, the shape is only subjected to pressure and no tension. Catenary shapes prove to be very useful when building with ice because it has very low tensile strength.

This inspired us to make an ice structure with a scale of 1/5 based on the original geometry of Gaudí's Sagrada Familia. The design consisted of one large dome, a nave and four entrance domes. The domes were constructed using an inflatable formwork as a mold, on which water, snow and Pykrete were alternately sprayed in thin layers to create a pykrete shell at a temperature of -8°C or lower. The columns of the nave were constructed with the same method by freezing a suspended rope structure.

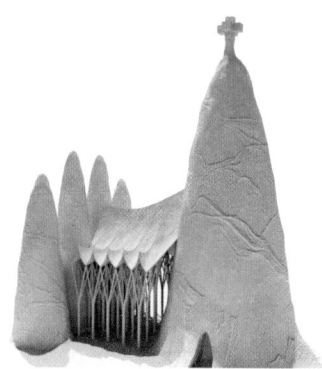

↑   Section and Render of the *Sagrada Familia in Ice.*

The form-finding of the domes is achieved with the software Oasys GSA. The primary geometry of the model was a parabola with a height of thirty meters and a width of 11.2 meters. By applying loads together with the self-weight of the shell, the program can form-find the ideal shape. The thickness of the dome has a gradient of 700 milimeters at the bottom to 300 milimeters at the top. The loadings on the ice structure are self-weight, snow loading, and wind load.

The *Sagrada Familia in Ice* was completed in a 24/7 process by four teams working in shifts. The water-cellulose mixture was sprayed on the higher parts and flowed downwards to the foundation. This gives the columns and shells a tapered section from the bottom to the top. Four of the domes (twenty-one and eighteen meters high) and the column structure of the nave were successfully constructed. Building with temperature-sensitive material creates structures that respond to and depend on their environment. Therefore the biggest dome could only be half realized due to the changing weather conditions.

↖ Spraying the Pykrete.

← Section through saw-dust-reinforced compos-ite ice material, Pykrete.

SAGRADA FAMILIA IN ICE

Fabric Formwork

Team

Prof. Mark West, *C.A.S.T.* Centre for Architectural Structures and Technology, University of Manitoba, & *Surviving logic*, artist studio, Montreal, Canada

Context

Built structures and drawings, 1981–present

Material

Concrete, polypropylene textile, rope, graphite drawings

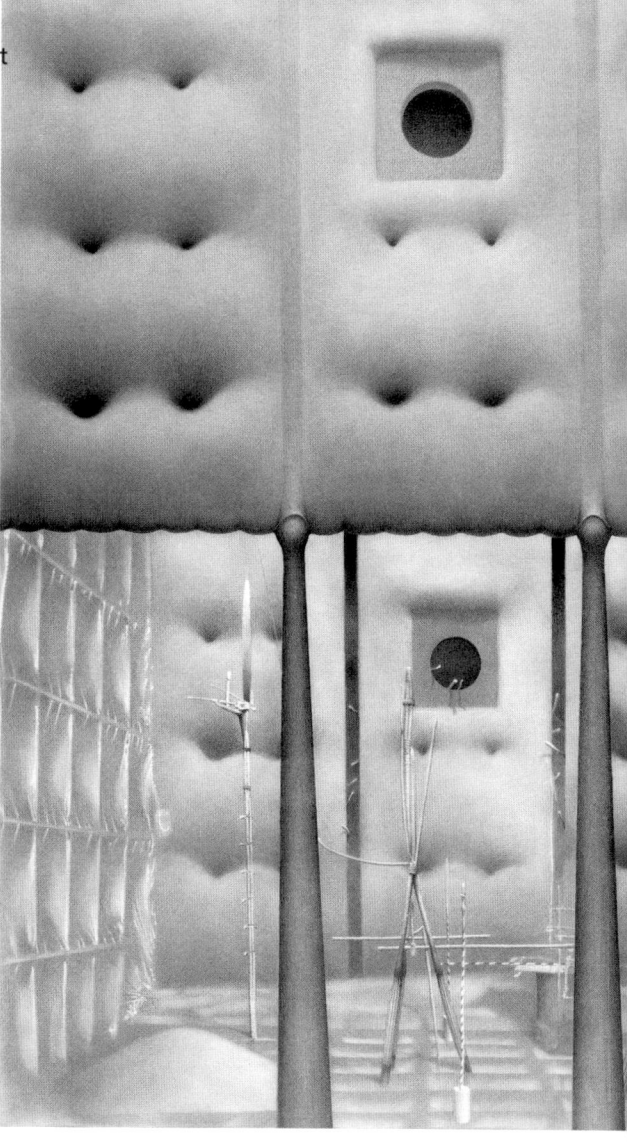

Mark West's approach to architecture reveals a pervasive textile materialism combined with a dimension of concrete that is not solid, but fluidly transitions into a soft, architectural expression. Capturing the moment when the wet cement mass slips into its textile skin, the process of making fabric formwork is deeply inscribed into the tectonics of this architecture. A shaping process takes place in which the soft formwork interacts with the natural hydrostatic pressure of the fluid concrete. The textile — which forms a flexible scaffold — is not only structural but also permeable. As the surface of the woven textile expands under pressure, the porosity of the fabric filters excess water, resulting in a stronger cast.

West's practice represents a way of understanding material as always vastly active, energetic, and organic. Transforming the visible pressure of the concrete into a taut or stiffened fabric, this particular technique allows the surface geometry to express itself, depending on how the formwork is arranged. The architect uses binding techniques that transform each seam, thread, and fold into unexpected shapes and bulges. The action of cutting and squeezing the textile gives a shape that is inscribed in the surface at the moment of casting. Because the technique is suitable for both on-site and precast use, the resulting qualities of fabric formwork are those of simplicity, participation, and material responsivity.[1] Compared to a traditional rigid wood or steel-framed formwork, less material is required for casting on site, which reduces waste. The naturally curved shapes always follow gravity and hydrostatic forces.

West pursues a sensual, spontaneous, and emergent approach to translating material from one form of artistic expression to the other. In his graphic drawings and years of sculptural work, West imagines architecture formed in a textile language and uses fabric formwork as a design method.[2]

1   M. West, *The Fabric Formwork Book: Methods for Building New Architectural and Structural Forms in Concrete* (London: Routledge, 2016).

2   "Entreentre Ee 1–interview. Drawings by Mark West, 2015," Entreentre, accessed April 20, 2022, https://entreentre.org/Ee01.html.

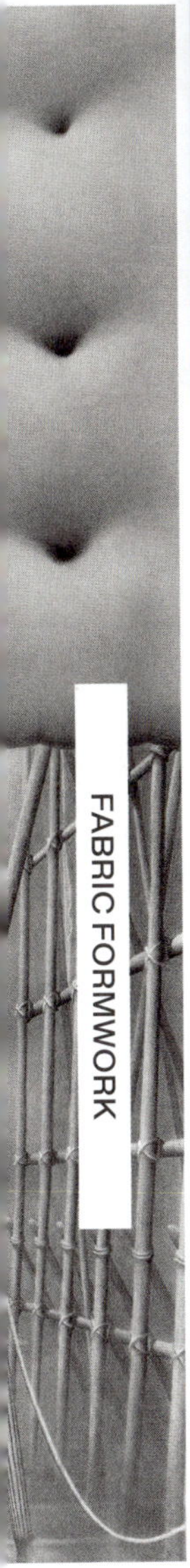

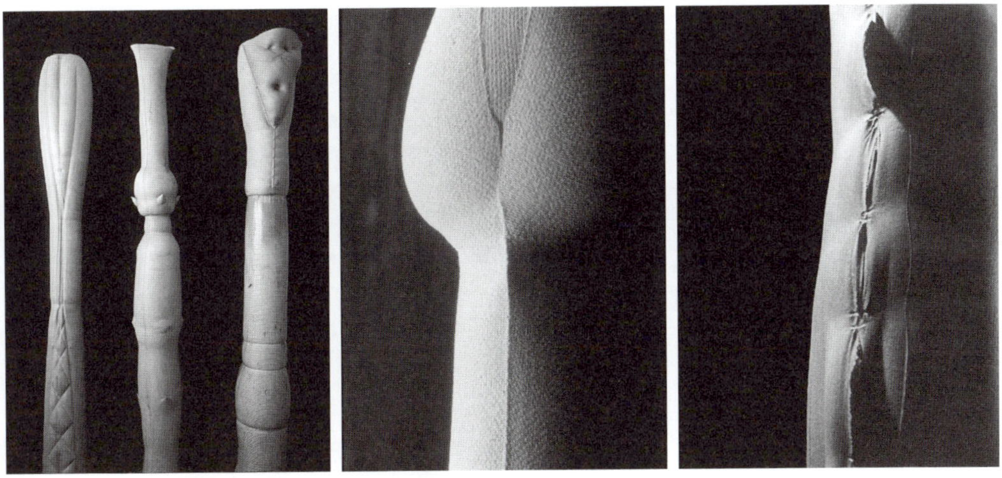

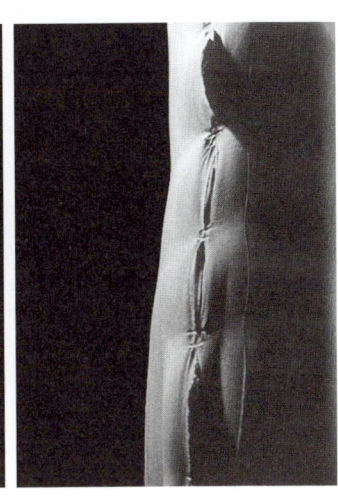

↑   Series of Small
Columns.

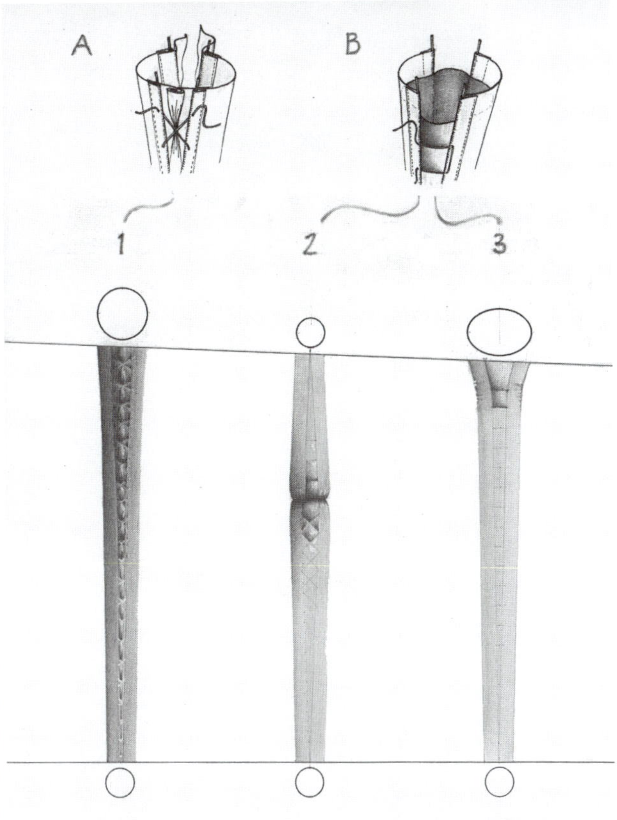

←← *Cracker House* drawing.

←   Proposal drawings for
Casa Dent Columns,
Culebra Puerto Rico.

↓   Fabric formwork for
concrete columns.

FABRIC FORMWORK

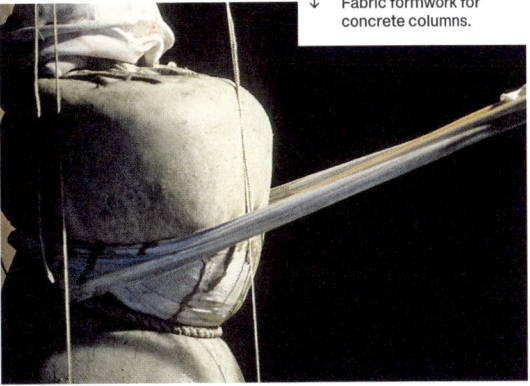

SunForceOceanLife

Team

Studio Ernesto Neto, Rio de Janeiro, Brazil

Context

Art installations:
SunForceOceanLife, The Museum of Fine Arts, Houston, USA, 2020
Aru Kuxipa | Sacred Secret, TBA 21—Thyssen-Bornemisza Art Contemporary, Vienna, Austria, 2015
In collaboration with Amazonian artists, plant healers and pajés (shamans) of the Huni-Kuin communities from the Jordão River area

Material

Crocheted textile, plastic balls (*SunForceOceanLife*)
Polyamide fabric, styrofoam, rice (*Aru Kuxipa | Sacred Secret*)

The immersive, soft environments of Ernesto Neto playfully explore concepts of tactility and gravity through the textile medium. Biomorphic landscapes invite visitors to enter and touch. Textiles are stretched across large volumes forming interior spaces to walk through or sit under. Elastic fabrics filled with granular materials such as rice or sand are connected and suspended from the ceiling, counterbalancing each other. Occasionally, the sculptural elements provide openings into which arms and legs can be inserted, or soft surfaces for bodies to sink into and lean against.

In *SunForceOceanLife*, a large spiral-shaped textile structure is suspended from the ceiling, hovering 3.5 meters above the floor. The net structure is hand-crocheted with textile chords in vibrant yellows, greens, and reds. The structural instability causes the spiral to slightly move and sway as visitors walk through it. The presence of other bodies is felt as each movement affects the overall balance—making one inevitably adjust to the motions of others.

Neto's focus on bodily, sensory experiences and how these individual forms of perception relate to those of others, is largely informed by the legacy of the political art of his native Brazil during the 1960s and 70s. In particular, his work has been influenced by the Neo-Concrete movement, with artists such as Lygia Pape and Lygia Clark, who sought to create participatory, multisensory spaces in which the impact of the work is felt through direct experience rather than symbolic representation. Neto continu-

↑ Installation view: Ernesto Neto, *SunForceOceanLife*, The Museum of Fine Arts, Houston, 2020.

→ Ernesto Neto and the Huni Kuin, *Aru Kuxipa | Sacred Secret*. TBA 21—Thyssen-Bornemisza Art Contemporary Collection, Vienna, 2015.

ously uses the tactility of textile materials to create sculptural landscapes that engage and address the body—altering how we interact with the built environment.

In the more recent installations, such as the above-mentioned *SunForceOceanLife*, the semi-translucent polyamide fabrics that characterized his early works have increasingly

been exchanged for handmade crocheted and knotted structures, shifting emphasis to textile craft and community building. In the collaborative works with artists, plant healers and shamans of the indigenous community of the Huni-Kuin, crocheted textile canopies and tent-like structures are used to transform the white rooms of the gallery space into intimate spaces for spiritual gatherings.

Fabricating Networks Quilt: Designing an Architecture of Social and Physical Spaces

Felecia Davis

INTRODUCTION

The artwork *Fabricating Networks Quilt: Stories from the Hill District in Pittsburgh Pennsylvania* is a quilt commissioned by the Museum of Modern Art in New York and exhibited as part of the *Reconstructions: Architecture and Blackness in America* show of spring 2021. A quilt is a multilayered textile made with woven cloth. Usually, a decorative top layer is made from patches of woven cloth, joined together with fancy stitching and backed by a plainer base layer of woven cloth. Between the two is a layer of batting.

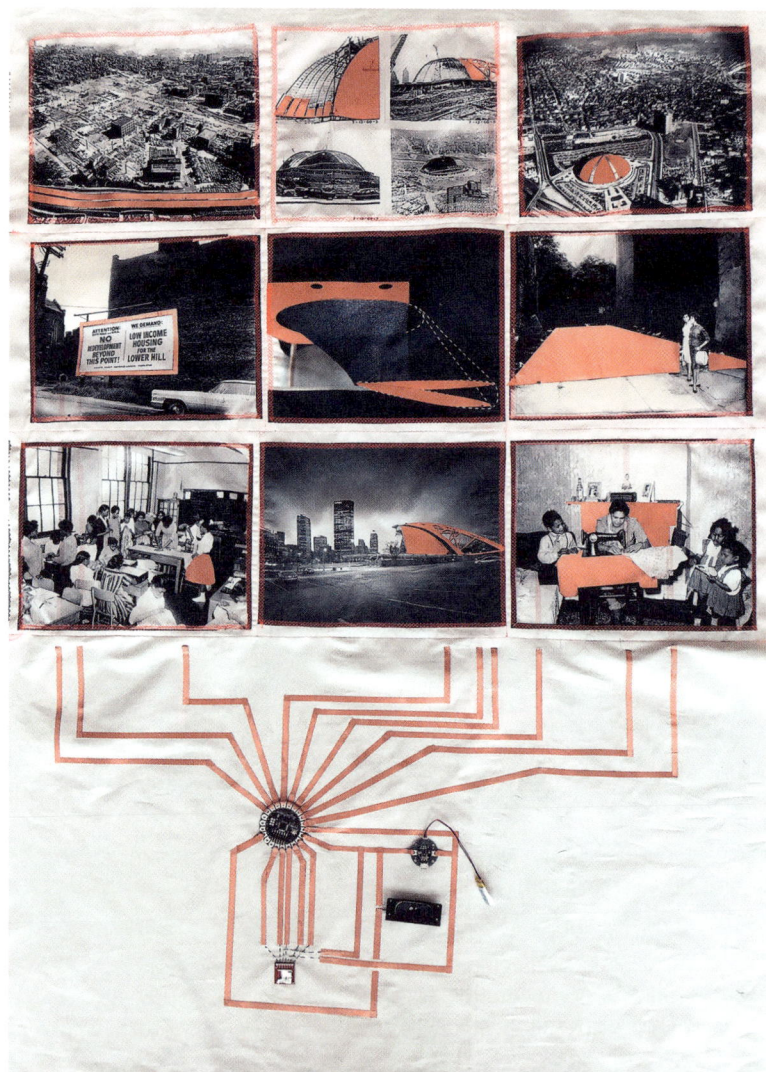

Fig. 1
Fabricating Networks Quilt (front). Scenes from the Hill District Pittsburgh. Image courtesy of Felecia Davis. Digital print on cotton broadcloth, copper-coated, ripstop nylon, copper tape, cotton thread, stainless-steel conductive thread, Lilypad microcontroller MP3 chip, speakers, battery.

Fabricating Networks Quilt operates as a computational textile made with fabrics and conductive textiles and fibers that are sensors connected to a microcontroller to communicate information. The makers of the quilt endeavored to connect with residents in a historically Black neighborhood in Pittsburgh to tell the stories of that neighborhood. Within the urban fabric the neighborhood showed the scars of the destruction wrought when making way for the interstate highway system and new commercial development. These are common scars that many Black neighborhoods across the United States still show and are the marks of the history of systemic dismantling and destruction of those neighborhoods. The discussion of the quilt moves from material craft and making to construction of a symbolic space of connection between architect/artist and the residents of the Hill District. The making of the quilt is a starting point to consider constructing a sustainable and physical architecture in Black neighborhoods.

THE EXHIBITION AND ITS GOALS

The exhibition *Reconstructions: Architecture and Blackness in America* was curated by Sean Anderson, Associate Curator at MoMA, and Mabel O. Wilson, Professor of Architecture and Professor in Africana Studies at Columbia University. They were assisted by Arièle Dionne Krosnick and later Anna Burkhardt, both Curatorial Assistants at MoMA. Opening in spring 2021, the exhibition was part of the recurring museum program *Issues in Contemporary Architecture* that focuses on topics with an urban dimension to bring forward concerns in contemporary architecture.

The focus of the *Reconstructions* exhibition was to address the unfinished project of reconstruction in the United States. Reconstruction took place in the brief time period between 1867 and 1877, after the US Civil War, when the government actively dedicated people, time and money to counteract the economic and social problems caused by the enslavement of Black people and tried to help shape their lives as new free citizens. Ever since, these freedoms have been challenged and curtailed, affecting every sector of life in the US.

In the words of the exhibition's curators:

This project investigates the intersections of architecture and anti-Black racism in the American context. The concept of the "racial" has been deployed to organize and conceptualize spaces of modernity, including those of museum exhibitions and collections. The exclusion of many Black builders and designers from architectural education and discourse has led to a severe lack of representation of Black architects in the architectural profession and related fields. Deliberate design decisions, along with federal, state and local policies and laws, have enabled and perpetuated *de jure* segregation, redlining, foreclosures, violence and mass incarceration, unequal access to public transportation, amongst other discriminatory practices in the urban sphere.[1]

The curators invited eleven artists to develop work in eleven different urban centers in the US, including Atlanta (Georgia), Miami (Florida), Syracuse (New York) Los Angeles (California), Pittsburgh (Pennsylvania), Nashville (Tennessee), Watts (Los Angeles), Oakland (California), Brooklyn (New York), New Orleans (Louisiana), and Kinloch (Missouri).

1   S. Anderson, M. O. Wilson, and A. Dionne-Krosnik, "Reconstructions," unpublished brief to artists, The Museum of Modern Art (2019), 2.

As one of those invited, I worked in the Hill District, Pittsburgh Pennsylvania, a historically Black neighborhood that bore the marks of racist design decisions.

THE HILL DISTRICT: CONSTRUCTING PLACE THROUGH PHOTOGRAPHS

The landscape of the Hill District, rising westward from the flat of Downtown Pittsburgh and its Cultural District, forms a separated geographic space that is less connected than expected to the city at large.[2] The Hill District rises westward from the flat of Downtown Pittsburgh and its Cultural District where the Allegheny and Monongahela Rivers meet. The Hill District is a thick, steep cord of land between the Allegheny and the Monongahela Rivers. It is flanked by two flats of land that line the banks of each river. This cord of land is cut off from the downtown by a steep rise and Interstate 579. The Hill District is most easily traversed along the length of the hill by its east-west avenues like Bedford, Wylie, Webster, and Centre Avenues. It is more difficult to move north and south up and over the hill, which takes one through a long, very steep greenbelt that is inaccessible to cars, and lines the hill before the flat.

This little hill/island in the city of Pittsburgh was the birthplace of Charles 'Teenie' Harris, a noted Black photographer who was the preeminent photographer for the *Pittsburgh Courier* between 1938 and 1975. *The Pittsburgh Courier,* a Black-owned newspaper, covered stories that addressed the concerns and perspectives of a Black audience. Teenie Harris' photographs showed the richness of everyday life in the city's Black neighborhoods of Pittsburgh. His photographs are often labeled with the street addresses and proper names of residents in the photos. Perhaps there are vacant lots with trash in the photographs, or broken sidewalks, but that is not the focus in his works. If one looks at the people in the photos, they are shaping and making their space. The focus is on people going about their daily lives. Harris' works are held in the Teenie Harris archives by the Carnegie Mellon Museum and form a foundation of work for connecting with Black life in the Hill District. I was drawn to this collection because these photos focused on stories and events that felt familiar to me. Many of the activities shown in Harris' photographs, for example the sewing club or sewing one's own clothes at home, in particular resonated. The photo of elegantly dressed women passing by the 1100 Wylie Street vacant lot reminded me of photos of my mother's family who grew up on the South Side of Chicago (Fig. 5).

THE QUILT AS RESEARCH AND CONNECTION

There are infinite stories that can be told about the history of life in the Hill District. The quilt was a start by a Black architect who lived outside of the community on the hill to connect with that community. The nine base panels of the quilt were fabricated by me and research assistants in my studio SOFTLAB@PSU. The quilt starts a conversation with people about making and is a repository of history. The idea of the quilt itself is not architecture but its act of making weaves together the history of the people and the place and becomes a method of questioning. The quilt idea was a way to start making sense of all the research that I was finding on Pittsburgh. A story common to many of the other urban centers unfolds in the

2   F. Davis, "Fabricating Networks: Transmissions and Receptions from Pittsburgh's Hill District," in *Reconstructions: Architecture and Blackness in America*, ed. S. Anderson and M. O. Wilson (New York: Museum of Modern Art), 54–59, here 55.

quilt panels. Knowing this history and understanding how people and families in the community are woven into the history of the US and the specific urban fabric is a necessary step to constructing anything new in neighborhoods that have been systematically destroyed.

The quilt is a computational textile made with ripstop copper-coated nylon, photographic digital print on cotton broadcloth, cotton thread, conductive thread, and a Lilypad microcontroller connected to a mini speaker and battery. When touched, the copper in each panel activates a speaker that tells a story of what happened in the photograph, taken from the archive. I would have liked to take this quilt on the road to the Hill District and continue to hear and make new stories and take new photographs with the residents on the Hill. It was meant to be an emotionally connected and networking material, but this was interrupted by COVID-19. The softness in this material system is how I defined the concept of construction of the communication which is collaborative and incrementally built. The Lilypad, designed by Leah Buechley, a designer and computer scientist at University of New Mexico, permits the fashioning of do-it-yourself work. It is supported by a network of online tutorials and programs. This network is also part of what we in my studio understand to be the *material*. For many of the works I do in the studio, the ability to access and use these inexpensive do-it-yourself components is the difference between connecting or not.

Fig. 2
Quilt panel showing urban renewal clearance for Civic Arena and I-579. Image courtesy of Felecia Davis. Digital print by John R. Schrader, "Lower Hill District clearance near Completion," 1957-58, Allegheny conference on community development, Detre Library & Archives Division, Senator John Heinz History Center Pittsburgh, PA.

Fig. 3
Quilt panel showing completed Civic Arena. Image courtesy of Felecia Davis. Digital print by Melvin Seidenberg (ca. 1828–1988), Urban Redevelopment Authority of Pittsburgh (Pa.), "Aerial Photograph of the Civic Arena and Surrounding Neighborhoods," 1961-71, Detre Library & Archives Division, Senator John Heinz History Center Pittsburgh, PA.

Fig. 4
Quilt panel showing future site of Freedom Corner. Image courtesy of Felecia Davis. Digital Print by Charles "Teenie" Harris (1908-1998), "Billboard inscribed 'Attention: City Hall and U.R.A. No Redevelopment Beyond This Point! We Demand Low Income Housing for the Lower Hill, C.C.H.D.R., N.A.A.C.P., Poor People's Campaign, Model Cities,' at Crawford Street near intersection of Centre Avenue, Hill District," 1969, black and white Kodak safety film, 4" × 5", Carnegie Museum of Art, Pittsburgh, Heinz Family Fund: © Carnegie Museum of Art, Charles "Teenie" Harris Archive.

Fig. 5
Quilt panel showing empty lot on Wylie Avenue, Hill District. Image courtesy of Felecia Davis. Digital Print by Charles "Teenie" Harris (1908-1998), "Two women, one holding hat box, standing in front of empty lot between two brick buildings with trash and rubble, 1100 block of Wylie Avenue, Hill District," July 1960, black-and-white Kodak safety film, 4" × 5", Carnegie Museum of Art, Pittsburgh, Heinz Family Fund: © Carnegie Museum of Art, Charles "Teenie" Harris Archive.

Fig. 6
Fabricating Networks Quilt detail: Lilypad microcontroller. Image courtesy of Felecia Davis.

The concepts of softness and openness – from thinking about the material to imagining and dreaming a new architectural process of physical construction – are important to the construction, materials, and stories of the quilt. These qualities are an idea found through heuristic consideration of the material qualities and operations happening in the quilt. Openness within order was a concept of the quilt that can be taken from thinking about the material to imagining and dreaming a new architectural process of physical construction. What are ways of imagining building like making a quilt?

Using quilt-making craft computationally, my graduate student, PhD candidate Farzaneh Oghazian, found a method to allow for the order of the quilt panels to be changed. The panels can be snapped together to read in a different order, thus making new stories or allowing people to understand new information.

An example of quilt-making craft can be seen in a project that Oghazian developed for her coursework. *The Circuit Game* aimed to teach children how to close a circuit using pieces connected with conductive copper ripstop nylon and stainless-steel coated fabrics, in order to light up a LED by tying groups of these together.[3]

3   F. Oghazian and F. Davis, "Circuit Game," in *Distributed, Ambient and Pervasive Interactions*, ed. N. Streitz and S. Konomi, HCII 2020. Lecture Notes in Computer Science, Vol. 12203 (Cham: Springer International Publishing, 2020), https://doi.org/10.1007/978-3-030-50344-4_45.

Fig. 7
The Circuit Game by Farzaneh Oghazian. Fiber composites, copper-coated ripstop nylon, conductive jersey knit, LEDs. Example of open and reconfigurable circuit. Image courtesy of Farzaneh Oghazian.

Fig. 8
Fabricating Networks Quilt
details, reverse side.
Connecting to the new Civic
Center (left) and the Sewing
Class (right). Image courtesy
of Felecia Davis.

CONTRIBUTIONS OF THE FABRICATING NETWORKS QUILT

The quilt and the fabrication of a social network is the preamble and an act of construction in a Black community that today lives with the scars of redlining and destruction through the use of seizure of the land by Eminent Domain for the purposes of new development. In the early 1960s, this development was the highway interstate system and the Civic Center. The same situation exists today. New development is slated for the empty hole left by the razing of the 1960s Civic Center that was no longer needed. Currently, the land at the heart of the Hill District where the old Civic Center sat is empty. It is now twenty-eight acres of parking lots waiting for another round of 'urban renewal'.

The intention of the quilt is to connect with residents in order to understand the past, to project a new present and desired future, constructed with community involvement. The quilt is a construction system that weaves together the present and past to make a social architecture that suggests a more sustainable process by building an infrastructure of people who together decide what is important and what dreams to pursue for their community.

ACKNOWLEDGEMENTS

The author would like to thank Mona Mirzaie for her help and leadership on the quilt, as well as Taylor Hufnagle, an intern in SOFTLAB@PSU. The project was supported by the Agnes Scollins Carey Memorial Professorship funds and the Stuckeman Center for Design Computing Innovation funds at the College of Arts and Architecture, Pennsylvania State University.

FURTHER READING:

S. S. Plummer, "Haptic Memory: Resituating Black Women's Lived Experiences in Fiber Art Narratives." (PhD diss., The Ohio State University, 2020).

Textile Fenestra and Spiracles: Mattering the swatch

Elaine Igoe

This is a nested exploration of rectangles of cloth. Swatches, bounded by architectonic framing, are permeable spaces within spaces. I propose that textiles offer sensorial fenestrations that facilitate the spiracular.

MATTERING

It is important to establish at the outset that textile swatches *are* designs. They commonly come forth from devices that afford them a rectangular shape—frames for making—whether looms, silk-screens, grids, sheets of paper, repeat units, computational screens, or windows. I begin by mattering the textile swatch or sample, considering this rectangular form through the architectonics of edges, surfaces, framing, and fenestration. Swatches tend to come in two forms. Firstly, as a designed thing in itself; in this case a swatch is a representation of possibility. Swatching and sampling is still the key commercial method used by creative textile studios and how most students learn textile design today. The second form of the swatch is one that has been detached from a large amount of existing material during or for the purposes of trading or communicating. These swatches are representations of plenty, as well as possibility curtailed, truncated, and controlled through the rectangular form. They are literally a "sample," a "taster." They exist and perform in rectangular formats in trade shows, material libraries, studios, workshops, factories, shops, and homes.

Rails upon rails, piles upon piles, rectangles of designs ready and waiting to take up their predestined role as one choice out of many. Multiplicitous in their compositions of colors, motifs, textures, fibers, and weights, they promise sensory delight in plentiful supply.

Fig. 1
Textile sample book from 1843. The Metropolitan Museum of Art, Gift of the Estate of Benjamin E. Marks, 1967.

Fig. 2
Buyer examining hanging
textile swatches at trade show,
2021.

In mattering the swatch, I underline the foundational, yet liminal position textiles hold in the chain of design. Designed as rectangular things, their shape changes in an entanglement with other designed things, from swatch or sample to dress, cushion, sofa, or drapes. Drawing on Henri Bergson, I am defining textiles as matter, viewed as an aggregate of images with an understanding that images have an existence halfway between a thing and a representation.[1] For me, this aptly describes the liminality of the textile swatch as the outcome of textile design. I believe that tacit framing practices feed back into the way we are designing textiles and designing with textiles. I do not here directly address textiles applied within architectural spaces, nor will I be concerned with structures of weaving. Instead, I will focus on what textiles can do once entangled with other designed outcomes or applied within a space involving aesthetics, tactility and function. Moving from edges, to surfaces, to framing and windows, to the affordance that a metaphoric window or fenestration provides, explored through notions of the virtual and spiracular, I will use architectonic constructs to understand the arguably hylomorphic nature of textiles.

EDGES

The edge of a swatch or sample of cloth is cut or self-edged / selvedged in the act of weaving or knitting. Edges of cloth are most usually hidden from view, stapled down, hemmed, concealed. The edges are the site where the liminality of textiles as design is activated. It is at the edges of cloth that we are reminded that they are a meshwork of lines; yarns, filaments and fibers that either turn back on themselves or are cut and perhaps fraying.

In my previous research I have addressed aspects of textile designing as translation. I use Gayatri Spivak's notion of *frayage*,[2] which means "facilitation," to understand the liminal positioning of the textile design. The fraying or undoing that, once cut, happens at the edge of a textile is an offering; a surrendering into a facilitation. That facilitation occurs in its application or in being worked into other designed things. The literal meshwork of the textile relies on this act of facilitation to continue to exist. Its facilitatory role is its key aim. As an assemblage, the structure delivers and surrenders

1   H. Bergson, *Matter and Memory*, trans. N. M. Paul and W. S. Palmer (New York: Zone [1908] 1988), 9.

2   G. Spivak, "The politics of translation," in *Outside In The Teaching Machine* (London, New York: Routledge, 1993), 179–200, here 180.

Fig. 3
The "active" edge illustrated in work by Erena Torizuka and Guy Genney from 2018, which combines woven Chirimen silk with computational sensors.

Fig. 4
The edge of a piece of Anni Albers' work (2017).

to facilitation, and this is aggregated by its aesthetic, sensory, and functional properties. And so, the significance and role of edges as framing devices for textile designing as the threshold between both the material and immaterial constituent parts must be explored. The movement between material and immaterial through the swatch operates apparition-like: it is not always clearly visible, particularly once a fabric has been further cut, sewn, digitally rendered, stapled, glued, and applied within the making and designing of other objects. Thus it is important to explore the transferable qualities of textiles before they are cut and sewn, and entangled in situ.[3]

3   A. Lottersberger, "Design, Innovation and Competitiveness in the Textile Industry: Upstream Design Innovation" (PhD diss., Politecnico di Milano, 2012), 46.

Ezio Manzini prompts us to begin a "rethinking of the role of a surface that emphasizes its character of autonomy from the rest of the object as well as the dynamic qualities that are concentrated in the surface."[4] He goes on to assert that the idea of static borders or edges of matter, in this context the edges of a piece of cut cloth, are replaced by the idea of the surface as an interface enabling an exchange of energy and information between the substances/media put into contact. He cites architect Andrea Branzi, who writes: "Unlike the surface of a painting ... a decorative surface implies infinite borders, and contains in each of its smallest parts the sum total of information in the entire system, since it contains the individual sign that is then repeated *ad infinitum*."[5]

Manzini also recognizes that repeated signs can create a rhythm.[6] Repetition and rhythm indeed remain one of the characteristic features of surfaces, whether the repetition of a weave or knit structure or in applied patterns. The various framing devices through which the activity of textile design and making happens, with their repetitious actions — spinning, weaving, stitching, knitting, printing, pasting, coding — translate creative processes into these rhythmic textures, motifs, and compositions that "go off the edge". These repetitious aspects serve as signifier of the creative action of the maker, as well as of the material-immaterial tension within textiles. Transgressing surfaces and edges challenges notions of framing and opens fenestrations.

FRAMING, WINDOWS, AND FENESTRATION

In *The Virtual Window*, Anne Friedberg quotes from Leon Battista Alberti's Renaissance treatise on painting and perspective, seeing the window as an aperture, opening and closing, separating the spaces of here and there. The window is also a membrane, "where surface meets depth."[7] Friedberg sees screens as "virtual windows" that change notions of materiality, space, and time. The screen is the latest rectangle to carry textile design activity to *frayage*. *Emancipath* is a collaboration between Zeitguised, a multi-disciplinary digital studio specializing in the creation of "exquisite realities," and Danish interior fabric manufacturer Kvadrat.[8] In these moving images, canvas becomes fluid, stretchy and bubbles, playing with our understanding of how textiles *are*. Textures are tiled and move across geometric shapes, intentionally exposing the work as a digital construct. The selvedges are moved to the center of the frame in an act of folding.[9] The capabilities of this immaterial cloth, though recognizable through a woven structure, are unknowable, unpredictable, and unreal. Post-digitality forces us to question known rules for dealing with tangible textiles, fabric grains running — literally moving and flowing — in opposing directions, cutting woven cloth to fit around 3D and moving shapes, weave structures that should not stretch, but do. In *Emancipath*, the edges instantiate these questions, becoming a *mise en abyme* of material and immaterial frames, edges, surfaces and windows.

4   E. Manzini, *The Material of Invention: Materials and Design* (Cambridge,
MA: MIT Press, 1989), 183.

5   Branzi 1984 cited in Manzini 1989, 196.
6   Manzini 1989, 192.
7   Alberti cited in A. Friedberg, *The Virtual Window: From Alberti to Microsoft*
(Cambridge, MA: MIT Press, 2006), 1.

8   Zeitguised, *Emancipath* (2017), https://zeitguised.com/emancipath.
9   G. Deleuze and F. Guattari, *A Thousand Plateaus: Capitalism and Schizophrenia*
(London: Continuum [1980] 2004).

Fig. 5
Film still of *Emancipath*
by Zeitguised, 2017.

VIRTUALITY

Returning to Alberti, the window is not a transparent window on the world but a "windowed elsewhere" — a virtual space that exists on the virtual plane of representation[10] — being understood by its classical root in the Latin *virtus*[11] meaning strength or power. In exploring virtuality, Friedberg reminds us that the virtual possesses a power of acting without the aid of matter: "[a]n immaterial proxy for the material."[12] In this sense, virtual imagery or representation, such like textiles being caught between thing and representation, has a "second-order materiality"[13] and a liminal immateriality. Friedberg uses the word "virtual" in Deleuzian terms, describing the virtual window as both a metaphoric window and an actual window with a virtual view held in place, the metaphor functioning as the point of

10   Alberti cited in Friedberg 2006, 243.
11   Ibid., 8.
12   Ibid.
13   Ibid., 11.

Fig. 6
Film still showing textile and
digital film works by Elena
Gučas, 2021.

transference into the virtual.[14] The virtual window, a framed surface with depth, becomes an opening, a *fenestra*, a portal to elsewhere. A textile swatch, just a small piece of cloth, performs as a virtual window with the power of acting as a proxy for the immaterial. Elena Gučas's work shown in Fig. 6 captures the nested action of framing, an action that crosses the threshold of virtuality. Here we see a textile swatch, suspended within a physical frame—framed in a filmic sense whilst juxtaposed by architectural framing devices of windows and doors. Gučas's montage of virtual windows facilitates, portal-like into an elsewhere of her lived experience.

SPIRACLES

The etymology of "swatch" in English uncovers a shift in meaning. Once the tally or tag attached to a piece of cloth to be dyed, it evolved to mean the detached sample piece of cloth itself. This linguistic link means that a swatch can be considered to have a counterpart elsewhere: it is a representative. Swatch and the term *swathe*, connoting

Fig. 7
Fabric token, chintz, 1759. A swatch left to identify a child at the Foundling Hospital in London, UK, which continues as the children's charity, Coram. The fabric was either provided by the mother or cut from the child's clothing by the Hospital's nurses. The swatches were attached to record books as proof of identity. Should the mother ever be able to return to their child, they would provide the counterpart to prove a match.

plentifulness, are both related to the noun *swath*. This is a space covered by the single cut of a scythe—a strip-like length—and subsequently highlights that space as a trace or vestige of an activity. It is at once an action, a thing and a space. This is poetically reminiscent of Simone Weil's "tear in a surface." A tear, brought about by the action of tearing, produces both a "thing" and a space. Elaine Scarry in *On Beauty and Being Just* talks about beautiful things that act like "small tears in the surface of the world that pull us through to some vaster space."[15] Absence here connotes possibilities and the

14 Ibid, 12.
15 S. Weil cited in E. Scarry, "On Beauty and Being Just," The Tanner Lectures on Human Values, delivered at Yale University 25-26 March 1998, https://tannerlectures.utah.edu/_resources/documents/a-to-z/s/scarry00.pdf, p. 77.

spectral presence of immanent, elsewhere counterparts. Thinking of textiles as fenestra, our visual and sensorial perception implicitly manifests the apparition of the immanent counterpart building this vaster space.

These portal-like concepts draw me to the notion of *the spiraculum eternitatis*, developed in the sixteenth century by alchemist Gerhard Dorn.[16] *Spiracle* is a biological term, a breathing hole in the anatomy of certain life forms. For Dorn, the *spiraculum eternitatis* is "a breathing hole into eternity." Just like "small tears," Dorn's spiracle is a window-like space into other realms. It is a threshold to alternative knowledge that can be manifested through functions of translation and art-making within tangible existence[17] — an effort to "translate that which is untranslatable." Translation, as mentioned earlier in reference to Spivak, can be understood as a function of designing.[18] Textile designers subjectively translate a narrative or mood through the selection and combination of materials, forms, symbols, colors, context, and function. The textile fenestra becomes a spiracle for a two-way translatory and communicative act between designer and viewer.

This is the operation of "textasis"[19] — the tension of "textility."[20] Textasis suggests an oscillation. Through the virtuality of the textile fenestra or spiracle the material element is in *stasis* — unmoving, subordinated, framed — while the immaterial is in *ekstasis* — plentiful, insubordinate, transgressing boundaries of edges and surfaces. The textile swatch is, then, a material–immaterial continuum occurring in "the flows of the currents of the lifeworld."[21] In this sense the rectangular swatch or sample of a textile can be understood as a framing device for translatory communication of embodied, enacted information. Textiles, even entangled in other designed things and spaces, hold us in spiracular textasis, a manifestation of an immanent elsewhere that we are invited to access.

TEXTILE FENESTRA AND SPIRACLES

16   D. Fremont, "The Spiracle in Alchemy and Art," *ARAS Connections: Image and Archetypes*, 3 (2017), 2, https://aras.org/sites/default/files/docs/00119Fremont.pdf 2.
17   von Franz cited in Fremont 2017, 2.
18   G. Baule and E. Caratti, *Design is Translation: The Translation Paradigm for the Culture of Design* (Milan: Franco Angeli Edizioni, 2017), 15.
19   E. Igoe, *Textile Design Theory in the Making* (London: Bloomsbury, 2021), 192.
20   V. Mitchell, "Textiles, Text, Techne" (1997), in *The Textile Reader*, ed. J. Hemmings (Oxford: Berg, 2012), 5-13; T. Ingold, "The Textility of Making," *Cambridge Journal of Economics* 34 (2010): 91-102.
21   T. Ingold, "Materials Against Materiality," *Archaeological Dialogues* 14, no. 1 (2007): 1-16, here 1.

Note: This text is based on "Where surface meets depth: virtuality in textile and material design," in *Surface Apparition: The Immateriality of Modern Surface,* ed. Yeseung Lee (London: Bloomsbury Academic, 2020).

Entangled Environments

Environments are entangled systems connecting scales and species. Constant change shapes their existence. The interdependency of their constituents creates networks in space and time. Obviously, biologically based material and design concepts are strongly rooted in this field, some of them showcased in this chapter.

Fibers are essential in such entanglements: mycelium filaments or plant roots build complex branching systems mostly hidden to the human eye. Recently, they have found their way onto the workbenches of architects, designers and artists. As examples, the case studies introduce mycelium that binds with wooden veneer into three-dimensional biocomposites, root structures that grow into sculptures, or biomimetic principles that inspire the design of foundations and coastal reinforcing.

In the animal world we can find fibrous environments built for specific purposes: birds weave sticks and grasses together to house their offspring, spiders spin silk into webs to catch their prey. Such "woven" architectures are highly complex and inspiring. When we engage in trans-species perspectives, we might listen to the vibrations of a spider string or redefine "waste" in our environment as new material for building.

With textile thinking, we may transform the walls of our homes from static barriers that shield the interior from allegedly hostile surroundings into a porous responsive fabric that embraces its environment and becomes part of a multi-scalar entangled metabolism where all elements will benefit from one another.

Biological Entanglements for Bioinspiration

Petra Gruber

Fig. 1
"Cathedral tree" in the tropical rainforest of French Guiana, around Cayenne.

"Weaving" can be defined in very technical terms, but for this essay I want to extend the topic to a metaphorical interweaving of disciplines such as architectural design and research in biology, physics, and engineering. Biomimetics, biomimicry and bioinspired design are terms used for the intersections of biology and technological design innovation.[1] Plants deliver intricate role models for structural design on many hierarchical levels, from the cellulose layered cell walls to the structure of a "cathedral tree."[2] In those trees, the ribs and hollow parts of the tree stem are "interwoven" in a macroscopic pattern, minimizing material and creating a structurally efficient lightweight tower (Fig. 1).

For this text I define weaving as generating ordered and functional structures through the combination of filamentous elements. The examples presented stem from entangled structures in biology and biomimicry research in the *BiodesignLab* at The University of Akron. They venture between exploratory design research and hypothesis-driven approaches.

BIOINSPIRED RESEARCH AND DESIGN

In Biology, fibrous elements are essential components of all organisms' tissues. Together with hierarchical layering, fibrous structure patterns are responsible for creating so-called emergent characteristics that deliver specific functionality in organismic design.[3] Plant biomechanics research has inspired new architectural products, technologies, and typologies (*Flectofin* mechanism, robotic filament

1   T. A. Lenau, A.-L. Metze, and T. Hesselberg, "Paradigms for Biologically Inspired Design," *Bioinspiration, Biomimetics, and Bioreplication VIII* 1059302 (March 2018): 1, https://doi.org/10.1117/12.2296560; P. Gruber, *Biomimetics in Architecture: Architecture of Life and Buildings* (Vienna: Ambra Verlag, 2010).
2   P. Gruber, "Biomimetics in Architecture—Inspiration from Plants," in *6th Plant Biomechanics Conference* (2009); T. Speck and I. Burgert, "Plant Stems: Functional Design and Mechanics," *Annual Review of Materials Research* 41, no. 1 (2011): 169-93, https://doi.org/10.1146/annurev-matsci-062910-100425.
3   U. G. K. Wegst, H. Bai, E. Saiz, A. P. Tomsia, and R. O. Ritchie, "Bioinspired Structural Materials," *Nature Materials* 14, no. 1 (2015): 23-36, https://doi.org/10.1038/nmat4089.

winding of fibrous modular systems).[4] Structures made by animals, such as birds' nests, spider webs and insects' cocoons also exhibit a purposeful combination of elements to structures that often include fibers (Fig. 2).[5] Climbing plants and plant-root systems are other examples of functional entangled biological architectures that explore environments in growth processes.[6]

GROWTH PATTERNS

In the arts-based research project *Biornametics*, the team around project lead Barbara Imhof and myself developed a bioinspired architectural installation that consisted of an intertwined technical fibrous structure and was installed in the covered courtyard at the University of Applied Arts in Vienna, Austria.[7] The follow-up arts-based research project *GrAB* (Growing As Building), allowed us to dive deeper into biological role models related to growth, and how biology can inform architecture and the arts. The project took growth patterns and dynamics from nature and applied them to architecture, with the goal of creating a new "living architecture." This 2.5-year research project allowed the setup of a biolab and looked at slime molds, algae, mycoterials, root systems, and many other organisms (Fig. 3). *GrAB* resulted in numerous concepts and prototype structures that were exhibited and published.[8]

BIOLOGICAL ENTANGLEMENTS

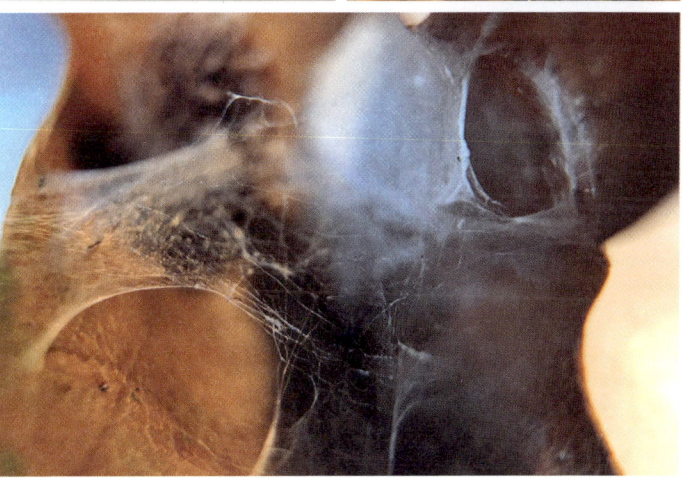

Fig. 2
Red Oak leaf with gull spun
into a spider cocoon, 2020.

4   J. Lienhard, S. Schleicher, S. Poppinga, T. Masselter, L. Müller, and J. Sartori, "Flectofin® A Hinge-Less Flapping Mechanism Inspired by Nature International," *International Bionic Award* 2012 (2012), 18; J. Knippers and K. G Nickel, *Biomimetic Research for Architecture and Building Construction* Vol. 8 (2016), https://doi.org/10.1007/978-3-319-46374-2.

5   M. Hansell, *Built by Animals: the natural history of animal architecture* (Oxford: Oxford University Press, 2007).

6   B. Mazzolai, L. Beccai, and V. Mattoli, "Plants as Model in Biomimetics and Biorobotics: New Perspectives," *Frontiers in Bioengineering and Biotechnology* 2 (January 2014): 1–5, https://doi.org/10.3389/fbioe.2014.00002.

7   B. Imhof and P. Gruber, *What Is the Architect Doing in the Jungle? Biornametics* (Vienna and New York: Edition Angewandte, 2013), https://doi.org/10.1007/978-3-7091-1529-9; P. Gruber and B. Imhof, "Patterns of Growth—Biomimetics and Architectural Design," *Buildings* 7, no. 2 (2017): 32, https://doi.org/10.3390/buildings7020032.

8   B. Imhof and P. Gruber, *Built to Grow. Blending Architecture and Biology* (Basel: Birkhäuser, Edition Angewandte, 2016).

NESTS AND BASKETS

The intricate connection between human cultural technology and biological construction was also explored in a design workshop held at the Bauhaus Bernau in 2017, where we researched the design of bird's nests and basket making from different European countries as a starting point for new design proposals. Nest-building in birds is very diverse, encompassing the large heaps of sticks made in tree forks by eagles and ospreys, to the elaborate cavernous structures

Fig. 4
Trash nest: collected trash compiled in analogy to scientific information on bird's nest building. Student project at workshop *Bauhaus Bernau*, 2017.

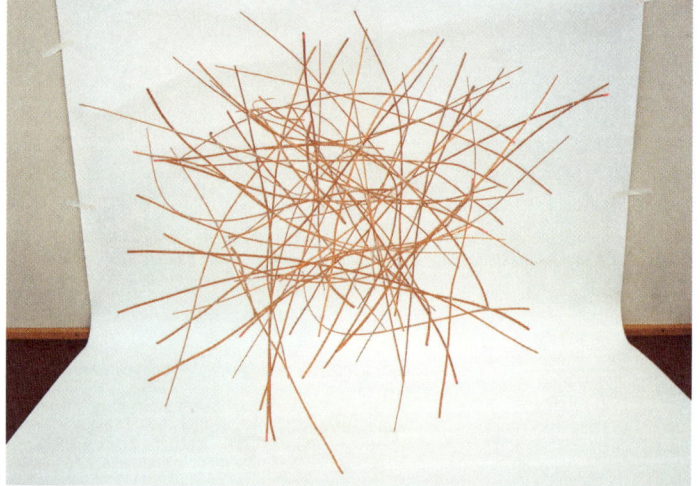

Fig. 5
Connectivity experiment: willow stick structure attempting to achieve a connectivity index of six (i.e. each element connecting with others at six contact points) with a random number of elements. Project at workshop *Bauhaus Bernau*, 2017.

of weaverbirds, to the composite fiber and mud constructions of American robins and many other smaller birds. In a workshop setting of one week, we prepared the group with scientific information on the physical characteristics and information on birds' nests as buildings.[9] Basic findings and principles of the layering of materials and mechanical performance in terms of friction, cohesion, compressibility were used to generate artistic designs. The trash nest, for example (Fig. 4) was not just a random assembly, but the result of a brief trash collection in the nearby woods, followed by sorting and combining according to the scientific findings on nest construction.

Another investigation researched the connectivity of stick nests in order to design sculptures (Fig. 5). This exploratory design research is inspirational for further design as well as basic research.

ROOT SYSTEMS

The ongoing research on "Biological root systems for biomimetic innovation for foundation design and coastal resilience" at the University of Akron, US, targets technical innovation in two distinct application areas, foundation and coastal engineering, that are prompted by climate change and ongoing severe weather events together with sea-level rising. The research, carried out by PhD students Elena Stachew and Thibaut Houette, is also inspired by the abundance of wetlands and forest landscapes of Northeast Ohio. In plants, the arrangement of cellulose fibers in the cell walls directly influences the mechanical properties at the respective body part. The overall stiffness and rigidity of the system is adapted by the radius and angle of the fibers, so a global shape is achieved by local parameters on a microscopic scale. Root systems of trees also exhibit great multifunctionality — starting with the mechanical properties of anchoring a massive cantilevering structure, they are responsible for nutrient and water uptake, exploration of the area for those elements, and as

Fig. 6
A bald cypress root was cleaned, marked with powder and positioned for photogrammetric imaging of the coarse root system. PhD students Elena Stachew and Thibaut Houette, The University of Akron, Davey Tree nursery, Ohio, 2020.

BIOLOGICAL ENTANGLEMENTS

9   L. Biddle, A. M. Goodman, and D. C. Deeming, "Construction Patterns of Birds' Nests Provide Insight into Nest-Building Behaviours," *PeerJ* 5 (2017): e3010, https://doi.org/10.7717/peerj.3010.

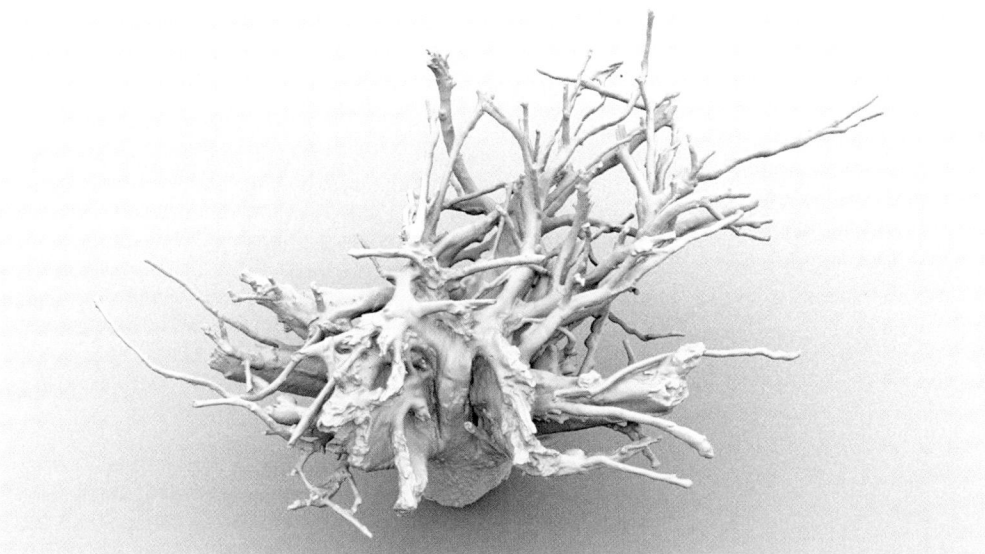

Fig. 7
Silver maple imaged with a
DSLR camera at Bath Nature
Preserve, Ohio, reconstructed
with Agisoft Metashape from
100 images, The University of
Akron BiodesignLab, 2020.

recent findings prove, communication with neighboring plants and soil organisms. Tree roots also prevent soil erosion, create a water-withholding system together with the soil, and serve as habitat, all important ecosystem services.[10] The analogies to architectural and engineering functions that were drawn and collected were published in a review paper.[11]

As generic as this may seem, establishing connections between fields is far from trivial and paves the way for further detailed research projects. Current foundation technology relies on the building mass and inertia to keep a building in place. Soil stability is required to withstand pressure and can be increased with depth or technical compaction. Heavy soil insertion techniques are applied, and a high safety factor is required due to lack of sensing and adaptability. Root systems could inform innovation in insertion techniques, more complex, better-connected branched morphologies, increased friction by surface extension and design, multifunctional use, adaptation to the environment and integration of sensing and control systems to create a dynamic response. "Self-x" functionality such as self-healing, self-assembly, self-cleaning, etc. would add time-based programmability to such systems.

We followed a biomimetic workflow, starting with grounding literature review and meta-research, followed by growth and morphology research, abstraction and transfer to technical solutions and application scenarios. Since current knowledge is merely descriptive and lacks 3D data, we generated virtual root-wad models using photogrammetry. By comparing different imaging and marker methods, scanning and available software programs, we developed a field methodology for the generation of 3D digital models. After a test-and-development phase in the lab and the collection of root examples from surrounding national parks, we applied the method strategically to three specimens of three different common tree species in the area. For the fieldwork on real trees, we collaborated with a nursery. The trees had to be removed with a large spade, and the root was cleaned, marked with powder, and positioned for imag-

10   J. E. Bidlack, S. H. Jansky, and K. R. Stern, "Chapter 5: Roots and Soils," in *Stern's Introductory Plant Biology,* 12th ed. (New York: McGraw-Hill, 2011), 622.
11   E. Stachew, T. Houette, and P. Gruber, "Root Systems Research for Bioinspired Resilient Design: A Concept Framework for Foundation and Coastal Engineering," *Frontiers in Robotics and AI* (2021), https://doi.org/10.3389/frobt.2021.548444.

ing (Fig. 6). Due to the process of cleaning and removal of fine roots, which cannot be imaged successfully with current technology, we were not able to keep the trees alive. Imaging was carried out with a high-quality camera and processed with the Agisoft Metashape program automatically into digital 3D models (Fig. 7). Several steps of data processing and cleanup were also necessary in the digital realm, and reconstruction of those complex models requires large computing power and time.

Currently, the 3D digital models are undergoing parametric analysis with Rhino and Grasshopper software. Topological and systems information will be abstracted and used to generate future foundation designs in an iterative process. A cycle of conceptual and parametric design, prototyping, experimental testing, learning, and modifying designs will optimize the proposals. According to the application scenario, multifunctionality aspects such as resource transport or filtering will be integrated.

Future work will include feasibility studies and finally lifecycle analyses of the new design proposals. Dynamic and multifunctional eco-system-serving building systems like this promise a very positive impact on the environment and the way we plan and build.

The integration of traditional craftsmanship such as weaving and scientific information from biology will without a doubt be helpful to develop more ecological and sustainable building technologies. More interdisciplinary research to understand the world around us is necessary, to leverage mutual beneficial effects into a new biohybrid design culture.

ACKNOWLEDGEMENTS

Thank you to all the individuals and institutions who contributed to the presented research. Special thanks go to: Co-project leader of Biornametics and *GrAB* Barbara Imhof; The University of Applied Arts Vienna, Austria, and the funding organization Austrian Science Fund FWF; the organizers of the Bauhaus Bernau Workshop and participating students; The University of Akron, Biomimicry Research and Innovation Center BRIC; The University of Akron Field Station and Davey Tree Company; PhD students Elena Stachew, Thibaut Houette and undergraduate students Claudia Naményi, Brandon Caster and Remik Niewiarowski.

Team
>Prof. Philipp Eversmann, Andrea Rossi, Nadja Nolte, Eda Özdemir, University of Kassel,
>Department of Experimental and Digital Design and Construction (EDEK)
>Prof. Dr. Jan Wurm, Ivan Acosta, Albert Dwan, Shibo Ren, Arup Deutschland GmbH
>Prof. Dirk Hebel, Dr. Alireza Javadian, Dr. Nazanin Saeidi, Karlsruhe Institute of Technology
>(KIT), Chair of Sustainable Construction

Context
>Research project funded by *Zukunft Bau*, 2021-23

Material
>Wood veneer, grown lignocellulosic mycelium
>Technology: wood welding, robotic placement, controlled bio-growth

Following the latest developments in the field of bio-based materials, the project *HOME* explores the structural enhancement of mycelium composites for applications in interior architecture and construction. With their carbon neutrality, lightness, fire-resistance and excellent sound-absorbing properties, these hybrid materials can be used for different purposes and be biologically composted at the end of their life cycle. They can take considerable compressive loads but have low tensile and flexural strength.[1]

Mycelium, as the root-like network of fungi, can colonize several square meters in a short time. It acts as a natural binder for biocomposites by growing a dense network of fine filaments. To gain a mycelium-based building material the fungal growth process is stopped by a heat treatment or drying process before the fungus develops fruiting bodies.

In order to guide the mycelium during its growth process, three-dimensional lattice structures from local wood species have been developed. The scaffolds serve as reinforcement and as mold; they are produced by a robotic wood-printing process that deposits thin wood-veneer strips to create spatial structures. During the deposition, the wood veneers require a binding system to keep their lattice form until the start of the mycelium growth. Therefore, ultrasonic welding, a technique that is common in various industries for fusing plastics or metals, was adopted for wood.[2] Welding

↑ Additive robotic manufacturing of 3D lattices with wood filament.

↓ 3D wood lattice made with solid maple veneer filament.

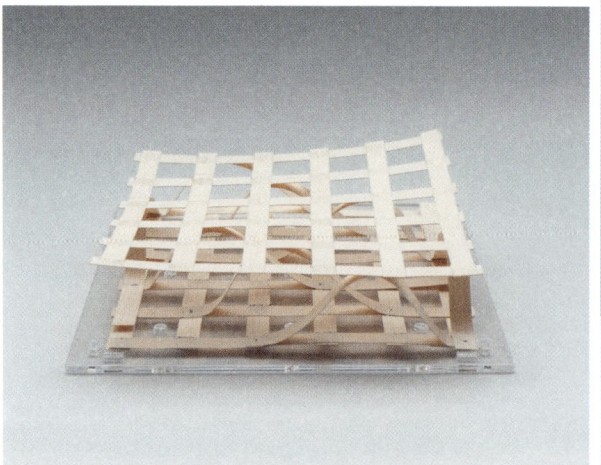

1   E. Özdemir, N. Saeidi, A. Javadian, A. Rossi, N. Nolte, S. Ren, A. Dwan, I. Acosta, D.E. Hebel, J. Wurm, and P. Eversmann, "Wood-Veneer-Reinforced Mycelium Composites for Sustainable Building Components," *Biomimetics* 7 (2022), 39, https://doi.org/10.3390/biomimetics7020039.

2   P. Eversmann, J. Ochs, J Heise. Z. Akbar, and S. Böhm, "Additive Timber Manufacturing: A Novel, Wood-Based Filament and Its Additive Robotic Fabrication Techniques for Large-Scale, Material-Efficient Construction," *3D Printing and Additive Manufacturing* (2021), http://doi.org/10.1089/3dp.2020.0356.

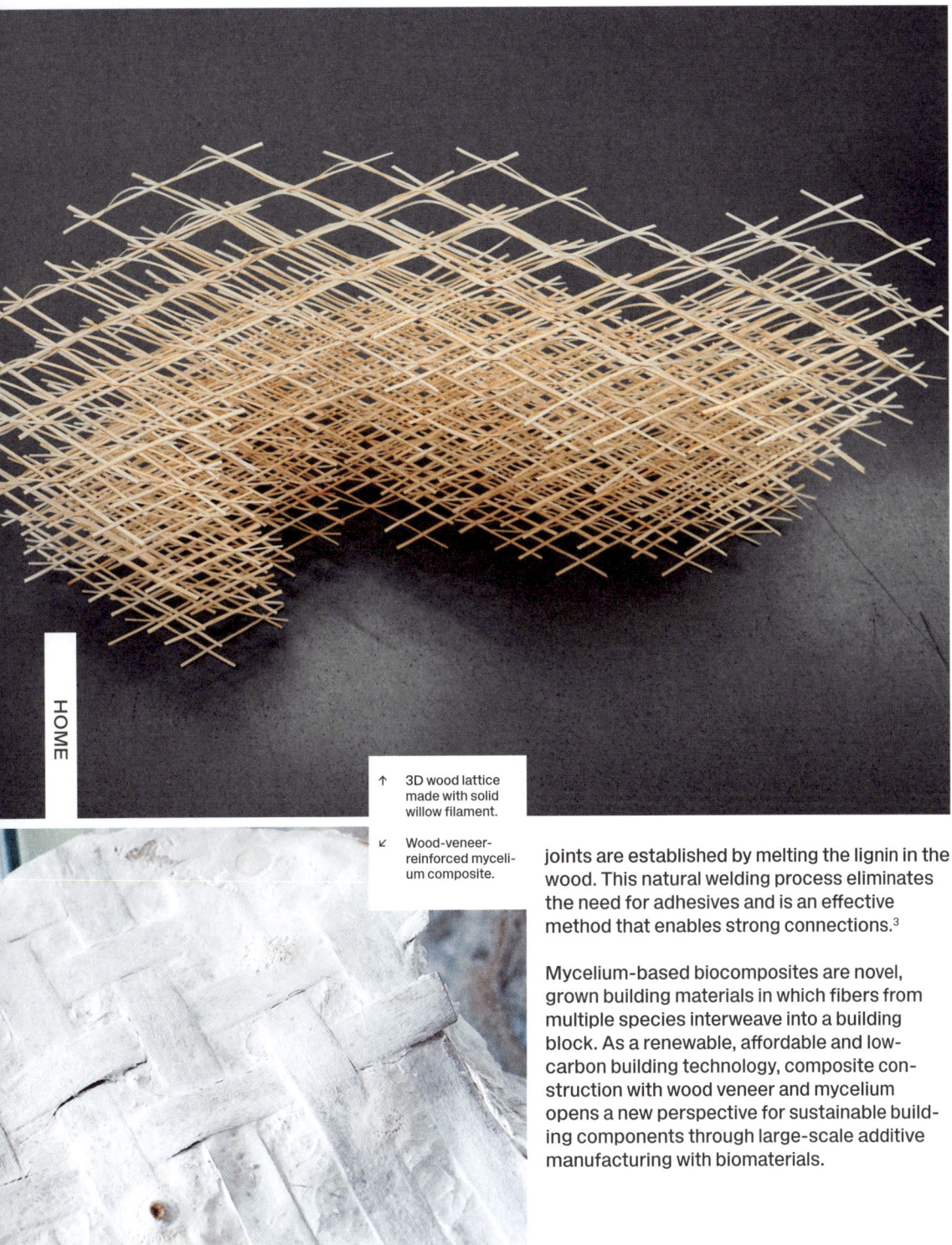

↑ 3D wood lattice made with solid willow filament.

↙ Wood-veneer-reinforced mycelium composite.

joints are established by melting the lignin in the wood. This natural welding process eliminates the need for adhesives and is an effective method that enables strong connections.[3]

Mycelium-based biocomposites are novel, grown building materials in which fibers from multiple species interweave into a building block. As a renewable, affordable and low-carbon building technology, composite construction with wood veneer and mycelium opens a new perspective for sustainable building components through large-scale additive manufacturing with biomaterials.

3   S. Silbermann, J. Heise, S. Böhm, P. Eversmann, and H. Klussmann, "Textile Tectonics for Wood Construction," in *Rethinking Wood: Future Dimensions of Timber Assembly*, ed. M. Hudert and S. Pfeiffer (Basel: Birkhäuser, 2019).

Interwoven—Exercises in Rootsystem Domestication

Team
    Diana Scherer, Amsterdam, Netherlands

Context
    Art project, artistic research, 2018-21

Material
    Various seeds, plant roots, soil, textile, photography

Working at the intersection of art, science, and design, visual artist Diana Scherer creates textile artworks with living botanical material by manipulating the otherwise hidden, underground networks of roots. Utilizing the dense root systems of field-growing crops such as wheat, oat, and grass, she creates intricate ornamental textile structures that reimagine traditional textile fabrication, blurring boundaries between the natural and the artificial.

After the seeds have been propagated in three-dimensional molds, the roots grow and intertwine in pre-defined geometric patterns. The technique was developed in collaboration with plant biologists from Radboud University in Nijmegen. Scherer's interventions in the growth process follow both scientific logic and intuition. The patterns are either derived from natural geometries, such as honeycomb hexagons, or from engineered patterns in her immediate surroundings like bubble wrap or metal floor grids. The plants are chosen according to aesthetic and haptic criteria, and the resulting dried material has a paper-like quality with varying characteristics depending on the plant, the thickness and the pattern structure of the mold. After approximately two weeks of growth underground, the works are "harvested." The result remains unpredictable each time. This interplay of control and letting go is an important element in Scherer's work.

Occasionally, the root structures are reinforced by placing loose textile filaments or open-weave fabric in the mold together with the seeds: the roots interweave with the fabric structure and form a growing plant root textile. The various stages of growth are captured and exhibited in large-scale photographs, which expose the fine details of the structures, such as the tiny hairs of the roots that help them intertwine.

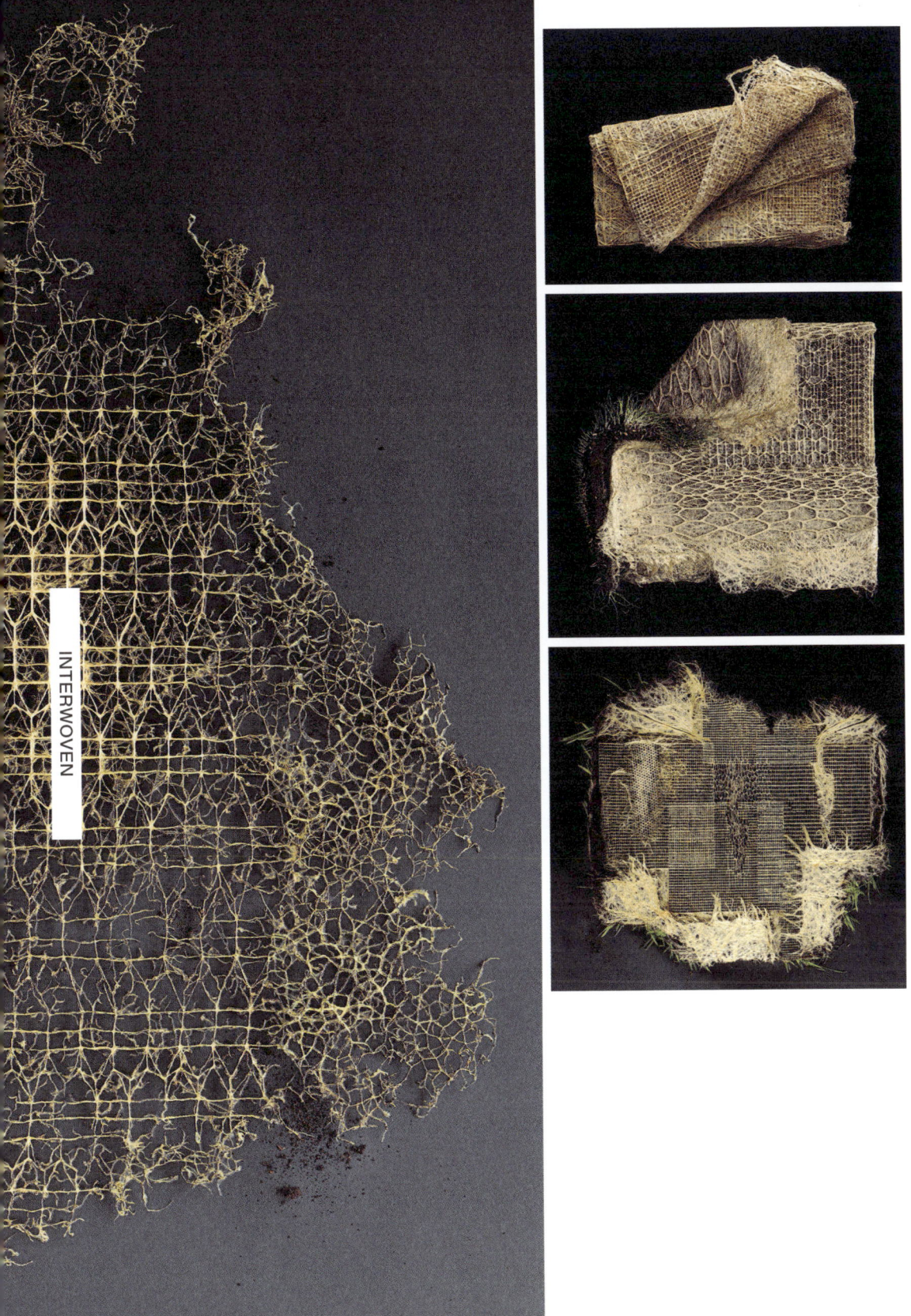

INTERWOVEN

Webs of At-tent(s)ion, Algo-r(h)i(y)thms

Team

    Tomás Saraceno, Berlin

Context

    Art projects, exhibition ON AIR, Tomás Saraceno, Palais de Tokyo, Paris, 2018

Material

    Webs of At-tent(s)ion:
    Seventy-six spider frames, spider silk, carbon fiber, lights, speakers.
    Spider silk provided by the species: *Nephila senegalensis, Cyrtophora citricola, Parawixia bistriata, Nephila edulis, Holocnemus pulchei, Nephila inaurata, Agelena labyrinthica, Psechrus jaegeri* and *Tegenaria domestica* among others

    Algo-r(h)i(y)thms:
    Rope, contact microphones, computer, machine-learning software, full-range active speakers, shakers, subwoofers, DMX-controlled light

The spider's web is more than simply a spatial architecture to the spider: it is an environment of vibrating strings through which it senses its surroundings. Like a spider web, the strings of the installation *Algo-r(h)i(y)thms* by Tomás Saraceno use vibration as a medium for communication, and they vibrate at different frequencies depending on their length and tension. Criss-crossing the space, the amplified strings form a sonic landscape that invites participants to create sounds that travel and merge with the vibrations created by others. The interconnected strings form clusters and nodes arranged according to spider web logic. The resulting installation is an interactive and collaborative musical instrument that the visitors enter.

For Tomás Saraceno the spider web is both a material metaphor for spatially extended perception as well as literally the material for his studio's art and research. In the installation series *Webs of At-tent(s)ion* different spider species, such as *Agelena labyrinthica* and *Cyrtophora citricola,* are brought together to weave their distinct spider web architectures in the same space. The intricate geometry of the various silk and web types form new entangled environments. Some of the delicate structures are amplified with highly sensitive microphones, making the rhythms of the spider vibrations audible to human ears.

Tomás Saraceno and his studio describe their work on spider webs as "speculative architectures" that aim to stimulate the imagination of interspecies attunement and co-existence.

Together with researchers from the Photogrammetric Institute of Technische Universität Darmstadt they have developed a tomographic technique that allows for precise three-dimensional scans to be made of complex spider

webs. The technique is not only a tool for creating large-scale art installations, but has also helped generate data that has in turn been used for further scientific research, as well as applications in architecture, engineering, and materials science. In the context of the global decline of spiders and other invertebrate populations, Tomás Saraceno and the Arachnophilia community seek to map the richness and diversity of spiders and their webs, proposing a new sensibility to the multispecies ecologies in which we are embedded.

←|↑ *Algo-r(h)i(y)thms*, 2018. Installation view at ON AIR, carte blanche exhibition to Tomás Saraceno, Palais de Tokyo, Paris, 2018. Curated by Rebecca Lamarche-Vadel.

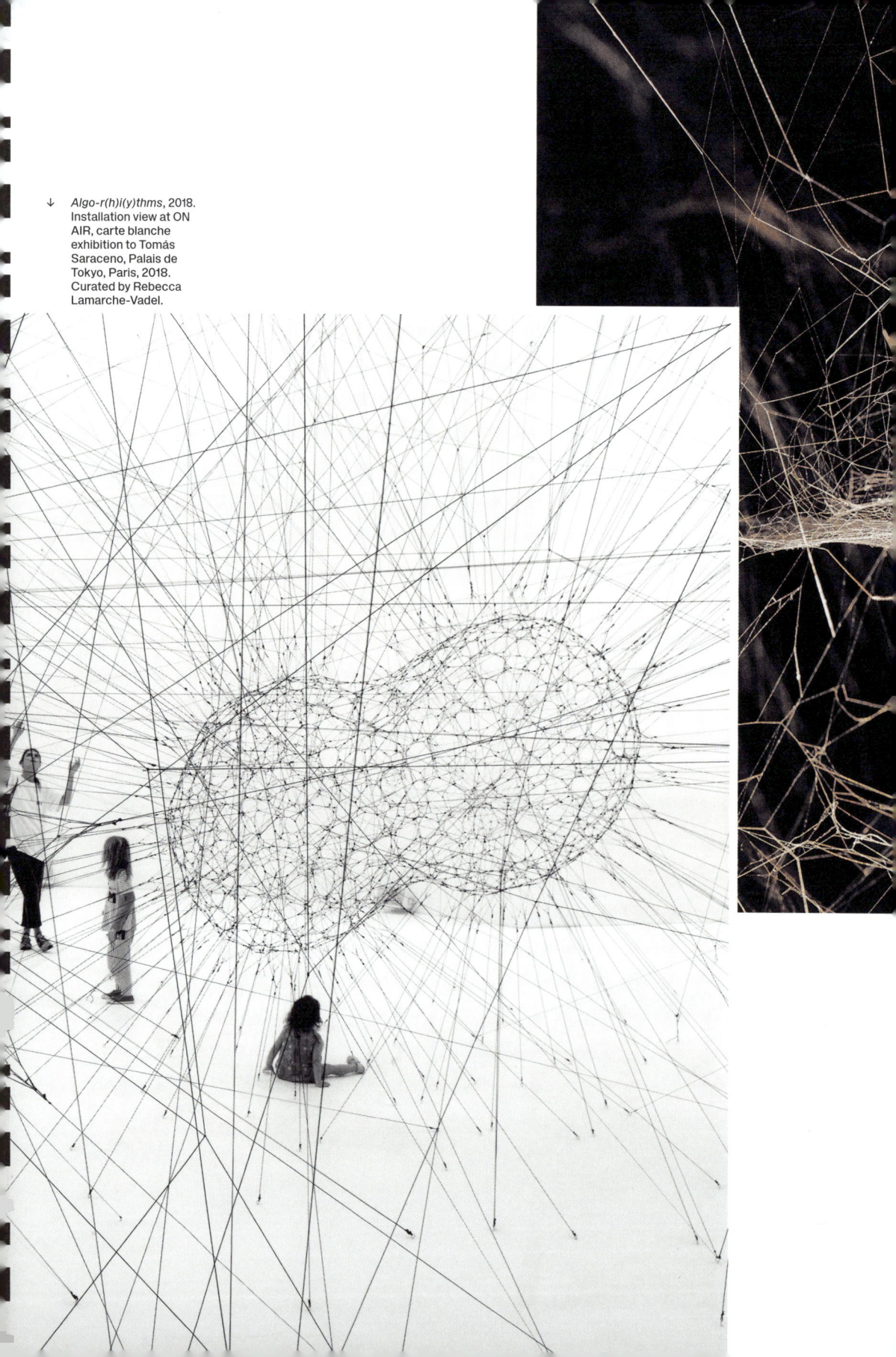

↓ *Algo-r(h)i(y)thms*, 2018.
Installation view at ON
AIR, carte blanche
exhibition to Tomás
Saraceno, Palais de
Tokyo, Paris, 2018.
Curated by Rebecca
Lamarche-Vadel.

WEBS OF AT-TENT(S)ION

↑ *Webs of At-tent(s)ion*,
2018.
Installation view at ON
AIR, carte blanche
exhibition to Tomás
Saraceno, Palais de
Tokyo, Paris, 2018.
Curated by Rebecca
Lamarche-Vadel.

Remarks on Designing for Multispecies Cohabitation—An Experimental Inquiry into the Biocolonization of Textile Facade

Svenja Keune, Delia Dumitrescu, Mette Ramsgaard Thomsen

New understandings in Western knowledge traditions about the "aliveness" of matter increasingly influence how we think about architecture and bring forth new expressions and properties of materials.[1] This paradigm shift in perceiving everything as dynamic and alive is promoted by the actual inclusion of living organisms, e.g. fungi, bacteria, vegetation, and wildlife, in the design of architectural structures, sustainable building materials, construction processes and life-cycles. Methods of design and production invite multispecies interactions and create new relationships between species and their environments. Urban spaces attract wildlife and invite us to design and build diverse forms of cohabitation between human and non-human species.[2] Dynamic qualities and aesthetics that are co-created by living organisms can challenge ideas and standards about, for example, the permanence of walls and their components such as cladding, insulation, or the outer facade. Therefore, they introduce alternative perspectives on architectural lifespan.

Architecture is increasingly shifting from designing, constructing, and materializing spaces for humans and human-centred activities to co-creating spaces with and for other species.[3] Hence, the needs of other species must be addressed and included in design processes at every level, as must their dynamic potential for co-creation as a driving force to creatively move design beyond anthropocentric approaches.

Making the needs of other than human species an integral part of design, planning processes, and architectural thinking opens up new symbiotic relationships between species, just as an ecosystem does. Different aspects of "living architecture" have been explored by several scholars: Rachel Armstrong[4] studied the transfer of resources through metabolisms by algae, while Shneel Malik et al.[5] investigated the biocolonization of facades by algae. Similar

1   M. Voyatzaki, *Architectural Materialisms: Nonhuman Creativity* (Edinburgh: Edinburgh University Press, 2018).

2   D. Goode, *Nature in Towns and Cities* (New York: HarperCollins Publishers, 2013); T. Donovan, *Feral Cities* (Chicago: Review Press, 2015).

3   B. Apfelbeck, R. P. H. Snep, T. E. Hauck, J. Ferguson, M. Holy, C. Jakoby, J. Scott MacIvor, L. Schär, M. Taylor, and W. W. Weisser, "Designing Wildlife-Inclusive Cities That Support Human-Animal Co-Existence," *Landscape and Urban Planning* (2020); M. Hensel, "The Rights to Ground: Integrating Human and Non-Human Perspectives in an Inclusive Approach to Sustainability," *Sustainable Development* 27 (2019): 245–51, https://doi.org/10.1002/sd.1883.

4   R. Armstrong, *Living Architecture: How Synthetic Biology Can Remake our Cities and Reshape our Lives* (Ted Conferences, 2012).

5   S. Malik, J. Hagopian, S. Mohite, C. Lintong, L. Stoffels, S. Giannakopoulos, R. Beckett et al., "Robotic Extrusion of Algae-Laden Hydrogels for Large-Scale Applications," *Global Challenges* 4, no. 1 (January 2020), https://doi.org/10.1002/gch2.201900064.

projects studied mosses and cryptogams,[6] insects,[7] artificial habitats for wildlife and "undesirable" forms of nature, e.g. puddles and migratory fowl.[8] In these projects, permeability and porosity are fundamental: Complex surfaces are designed to incorporate crinkles and folds in order to provide space for the accumulation of water and bio-colonization of all forms of life. Research in the context of architecture has thus far primarily focused on exploring bio-colonization through rigid architectural materials like concrete, soil,[9] and plastics.[10]

We are particularly interested in exploring the design potentials that emerge when combining stiff and flexible materials, e.g. textiles, to enable multi-species cohabitation. Hence, the approach that was used in the research presented in this chapter was to explore the design and usage of porous textile hybrids for negotiating space with living organisms. We focused on woven textiles with biopolymer prints and examined how surfaces featuring soft and rigid qualities as well as voids and pores could be made habitable for insects and plants in order to create shared, osmotic membranes in which human and non-human environments can connect.

The project began with the construction of a textile framework: cotton-gauze fabrics served as flexible screens onto which a biopolymer paste was applied. The soil-based paste was manually extruded to create porous, three-dimensional structures with large surface areas that would enable different conditions for the attachment, growth, and inhabitation of microorganisms, plants, and insects.

The resulting hybrid textiles were framed and hung vertically in an experimental house extension built onto the Tiny House on Wheels parked in Hvalsø, Denmark. They were protected from direct rain but otherwise exposed to the outside and the temperate climate. The aim was to investigate how these textiles negotiate spaces with spiders, solitary bees, and other types of insect, plants, and microorganisms. This facilitated close cohabitation and observation of the experiment at different times of day and across the seasons. Our strategy was not to deliberately design for a specific species, but to offer an open invitation by creating textile hybrids: collages of different materials such as soils and twigs, organized in various patterns in order to achieve different visual, haptic, olfactory, and gustatory expressions.

A flexible textile base supported different forms of layering of biopolymers derived from local soils on one or both sides (Fig. 1, 2). The flexible textile substrate was an important carrier of the printed pattern; it facilitated the shape- and state-changing qualities of the soil- and sawdust-based biopolymers, which were rigid and stable in a dry atmosphere and became more flexible and gel-like with increasing moisture due to humidity or rain. The hybrid textile was therefore reshaped and re-draped numerous times. It was speculated that the biopolymers can serve as a framework for the growth of microorganisms in moist conditions, while a dry atmosphere provides stability and an environment in which insects could find shelter, materials to build on and space to breed.

6   M. Cruz and R. Beckett, "Bioreceptive Design: A Novel Approach to Biodigital Materiality," *Architectural Research Quarterly* (2016), https://doi.org/10.1017/S1359135516000130.
7   M. Joachim, M. Aiolova, and O N E Terreform, *Design with Life: Biotech Architecture and Resilient Cities* (New York: Actar, 2020).
8   D. Gissen, *Subnature: Architecture's Other Environments: Atmospheres, Matter, Life* (New York: Princeton Architectural Press, 2009).
9   D. Mitterberger and T. Derme, "Soil 3D Printing Combining Robotic Binder-Jetting Processes with Organic Composites for Biodegradable Soil Structures," *Ubiquity and Autonomy — Paper Proceedings of the 39th Annual Conference of the Association for Computer Aided Design in Architecture* (ACADIA, 2019).
10   C. Pasquero and M. Poletto, "Beauty as Ecological Intelligence: Bio-Digital Aesthetics as a Value System of Post-Anthropocene Architecture," *Architectural Design* (2019), https://doi.org/10.1002/ad.2480.

Fig. 1
A close-up of a one-sided
single-layered circular pattern
with two biopolymers, the dark
brown based on soil, the bright
brown based on sawdust.

Due to the high (up to 70%) water content of the biopolymers, an internal porosity was created in the biopolymer print when they dried. The process of evaporation also caused the contraction of the manually printed structure, pulling the textile substrate into a new form. There was a co-dependency between the textile and the biopolymers: in a slurry state, the textile provided a foundation for the printed biopolymers; in a dry state, the biopolymers reinforced the textile, giving it stability and holding it in a certain shape. This shape, however, could be manipulated as the biopolymers reverted to their slurry state in response to exposure to moisture. Therefore, the textile hybrid changed state and shape numerous times and was able to recover from cracking.

The textile biohybrids that were explored in this research could provide an alternative perspective on facades and multispecies cohabitation. The textile bears the print and facilitates dynamic expressions by allowing the shape- and state-changing biopolymers to form patterns. It is also easily manipulated through coloring applications, adding three-dimensionality through surface manipulation, and forming through draping. The flexible textile base made the handling and set up of the hybrid textiles easier, expanding the ways in which they can be set up and arranged. The textile also serves as a filter or physical barrier, mediating the access of organisms through the weave structure, as shown in Fig. 4. The biopolymers served as sources of nutrients, building materials for use by multiple species, reinforcement, a shaping agent and adhesive to bind other materials such as seeds and twigs to the textile. They also added three-dimensionality and porosity, providing the potential for growth and life in terms of plants and insects.

Fig. 2
A one-sided punctual organization of a soil-based biopolymer with attached pieces of hollow nettle stalks, which allows the fabric to express its flexibility and the stiff elements to follow the fabric's dynamic expressions.

The experimental structure of the research enabled us to engage in two levels of understanding: the first level was constituted by the experience of making the textile hybrids, and the second level was enacted by experiencing the textile hybrids in use. The intertwining of the two research approaches allowed us to exemplify and speculate on the possible design futures of these hybrids from multiple design perspectives, including those of the biopolymer, the textile, and the living world. Here, biodegradable materials such as cotton, soil, and sawdust were exposed to environmental factors as a deliberate way of orchestrating the temporal performance of architectural facade elements. Moreover, the layered textile facade provided a complementary perspective on the question of temporality and duration in relation to the lifespans of both architecture and its occupants. As the textile hybrids were occupied, they were gently changed by their dwellers; a caterpillar spun itself a cocoon, another explored the nettle stalks applied to a biopolymer print on a textile, spiders lived and hunted in the spaces between textile folds and printed pattern, and mold covered a white cloth in a pattern of small dark dots. As such, the organisms became part of the changing aesthetics and dynamic processes of textile weathering and erosion

Fig. 5
A textile hybrid explored by
a moth.

in relation to an architectural facade. By changing in this way, the established norms in the architectural design practice that are famously founded on ideas of permanence and stasis[11] are challenged. The deliberately designed natural growth and decay challenged the aesthetics and expressions of the textile hybrid in relation to the lifespans of its living inhabitants. As a work in progress, the design probes examined how the design of porosity and aperture can steer erosion, creating the basis for a novel sympoetic aesthetic language for architectural facades, moving the design perspective from human-centered to multispecies cohabitation.

The complex constituency of these contemporary architectural "skins," i.e. textile facades, requires a new systematic approach to the design and planning of material assemblies and their life spans. Here, the character of the facade depends on the constituent layers of the textiles, dwellers, and the added layers of biopolymers, which support the growth of plants and interaction with insects. In addition, this complex material assembly opens for exploratory craftsmanship that mixes the knowledge of different material domains such as textile constructions and additive methods for biopolymer printing when designing the hybrid textile system.

Nowadays, sustainable thinking in architectural design allows new perspectives on the use of textiles in building design with regard to the building's longevity, environmental impact, and the possibility to design with various materials' life spans, ranging from durable polyester polymers to highly degradable bio-based ones. These extended possibilities found in textiles come to support current architectural restoration theory that conceptualizes the temporality of material systems as part of a sustainable building culture by borrowing the notion of ecosystem hierarchy and its embodiment

11   M. Taylor, "Time Matters: Transition and Transformation in Architecture," *Architectural Design* (2016), https://doi.org/10.1002/ad.2000.

Fig. 6
The Tiny House with its extension in which the textile hybrids are set up as facade elements.

on different timescales.[12] The experimental inquiry presented in this chapter focuses on the development of such temporal systems for the biocolonization of textile facades and the creation of dynamic forms of cohabitation that emerge between various insects, plants, microorganisms, and humans. Likewise, this experimental approach aims to expand the understanding of "alive" matter and sustainable architecture in relation to embedding differentiated functions and the natural uses and lifespans of its inhabitants; it speculates on the creation of interconnected textile systems in which human and more-than-human entities can be harmoniously interrelated through architectural design.

12   R. O'Neill, D. L. Deangelis, J.B. Waide, and T. F. H. Allen, *A Hierarchical Concept Of Ecosystems* (Princeton: Princeton University Press, 1986).

MULTISPECIES COHABITATION

Fig. 7
A caterpillar exploring a
curtain suspended outside the
greenhouse (the image refers
to Keune's previous project *On
Textile Farming*).

Living with Moths

Jörg Petruschat

FIG. 11. Plan of building-level VII.   *To face p. 52*

Fig. 1
Çatal Höyük building-level
VII, ca. 7000 B.C.E., from
J. Mellaart, *Çatal Hüyük: A
Neolithic Town in Anatolia*
(London: Thames and Hudson,
1967).

Architecture emerges as a practice of demarcation: for a social existence, interior spaces are separated from exterior spaces. The space within architecture allows for an awareness that the world outside may be an oppressive one, but that it is not the only reality. The spatial seclusion permits the implementation of mental models: possible realities besides the one outside are simulated. The experience of the real is joined by the imagination of the virtual.

Before one of the oldest settlements — Çatal Höyük — was built in the Neolithic period, the site was marked by an arrangement of sacred stones. Later, one of the largest settlements in the early period of mankind was built nearby, not as an accumulation of solitary buildings, but as a formation of housing units following the pattern of cellular growth (Fig. 1, 2).

The architectural act of excluding and enclosing is based on modalities of the biological. In order to build up metabolism and information exchange, something we call growth, organisms need to segregate their biochemical processes from earthly entropy: membranes, skins, shields offer the possibility for an organismic order of the inside. It is fascinating to think that these existence-generating modes of enclosing and excluding lead to life in cells as well as to the emergence of architecture. The earliest temples of the Near East delimit a walled area for the ritual action of priests by separating it from the existence of all else. This sacred area is initially without a roof, maintaining the visible connection to the heavenly. Only by adding the roof is the closing-off from the environment completed. Then the immediate connection to the cosmic is suspended and the access to the almighty becomes a matter of evocations and simu-

Fig. 2
Çatal Höyük: wall painting in a tomb underneath one of the buildings, ca. 6000 B.C.E., from *Sci News*, January 13, 2014. Neolithic mural in Çatalhöyük, Turkey, and its interpretive drawing by Grace Huxtable, in J. Mellaart, *Çatal Hüyük: A Neolithic Town in Anatolia* (London: Thames and Hudson, 1967).

lations. From now on, the cosmic orientation of human existence becomes a matter of architectural layouts.

In both these origins of architecture named above—the residential structures of early settlements following the pattern of cell structures as well as the enclosure of a sacred area—the wall is the constitutive element. It is spatial conclusion, inclusion, exclusion. With the wall and its motif of demarcation, support and load come into play. Demarcating, supporting and loading are signatures of settling down, dependent on the gradient of gravity.

During the French Revolution, the architect Claude-Nicolas Ledoux started the project of the Ideal City of Chaux, drawing a house for the gardeners in the shape of a sphere (Fig. 3). At least visually, this building has abandoned any anchorage to the site, although it is massively walled. It is an architecture freed from resting on the earth that imagines the cosmic dimension of human existence (Fig. 4). The evolution of architecture does not follow this path. It does not give up the bond to earth and gravity. The use of steel dissolves walls, puts floors and roofs on ever narrower supports. It eventually hangs a curtain of glass in front of the load-bearing structure, so visually at least the inside affects the outside and vice versa. With the façade hung in front of the load-bearing structure, the juxtaposition of inside and outside demanded by the concept of the wall is eliminated. The forces in the curtain wall are forces of tension and no longer of compression, as in the wall rising from the ground. The curtain appears like a lattice with cladding elements simply hung inside it. This justifies the term "curtain."

Curtain walls can open and close much more efficiently than structural walls ever could. Although walls always contain openings that can be closed, these must be freed from the compressive forces of the wall with the help of distinct supports, arches, beams. For curtain facades, opening and closing is an unencumbered structural quality.

In fact, the classical concept of the wall is overcome in the evolution of architecture by tying it to the constructive potential of the roof. For as long as there have been walls, the roof has been the cover that suspended the free view to the sky, protected inhabitants from a hostile climate, vaulted the interior and enclosed the social space below. The aesthetic and climatic closure from above offered an immersive quality to simulations of realness beyond the real world. Without the ceiling, there would be none of the overwhelming simulations of otherworldly scenarios in Renaissance and Baroque.

From the perspective of construction, the roof is, above all else, a surface hanging down on itself. This was already the case in nomadic times, both in the tent and in the yurt. From the very beginning, the supporting structure of tent and yurt was separated from the covering surfaces that protect the inside from the outside. The scaffolding was erected first, then the surface screens for roofing were tensioned or thrown over and tied down as a protective wall. This structural separation of load-bearing construction and cladding returns in the concept of the curtain wall.

Fig. 3
Claude-Nicolas Ledoux,
Maison des Gardes Agricoles
(Garden House, Design for
the Ideal City of Chaux), from
C.-L. Ledoux, *L'architecture
considérée sous le rapport
de l'art, des moeurs et de la
législation* (Paris, 1804).

At the Federal Garden Show in Kassel in 1955, the architect Frei Otto erected a music pavilion designed as a textile roof stretched over four points (Fig. 5). A tent company had supplied him with the material and know-how. A year earlier, Otto had written his doctoral thesis on the subject of "The Hanging Roof." He had been a glider pilot from an early age and therefore was well-acquainted

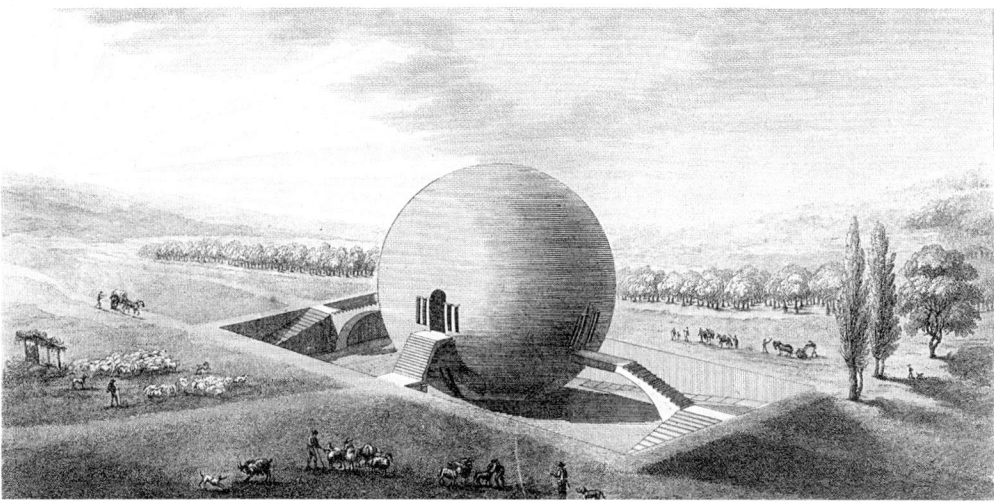

ÉLEVATION DU CIMETIÈRE DE LA VILLE DE CHAUX.

Fig. 4
Claude-Nicolas Ledoux:
Élevation du Cimetière de la
Ville de Chaux (Elevation of
the Cemetery Buildings of
the Ideal City of Chaux), from
C.-L. Ledoux, *L'architecture
considérée sous le rapport
de l'art, des moeurs et de la
legislation* (Paris, 1804).

with membranes stretched over frameworks. Textile appeared suitable to him for taking up the forces of a tensioned construction. He developed the concept of his tent roofs neither from weaving nor from camping, but from the geometry of minimal surfaces. For Otto, textile is not a design medium with distinct qualities, but a device for staging surface structures. His inspirations derive from the observation of soap bubbles rather than from the exploration of fabrics and their structural interpretation.

Fig. 5
Frei Otto, Music Pavilion for the Federal Garden Show in Kassel, 1955.

In order to unfold the potential of an architecture that originates from fabric, the imagination must transcend the classical archetypes of wall, ceiling, roof, and the motif of demarcation between a here *inside* and a there *outside*. It must also leave behind the hitherto existing, culturally limited metaphors of textile as clothing and cladding that Gottfried Semper had introduced into the conceptual discussion. The architectural invention must detach itself from the anthropocentric view in the dimensions of human clothing, in which the garment is only a separating layer that may have folds, pockets, and a symbolic decorum, but does not attain a spatial quality of its own.

The first and foremost potential of fabrics is that they allow for a fundamentally different understanding of architecture: instead of conceiving architecture as a detachment and separation of human spaces from all other existence, fabric permits one to identify a distinct space of encounter and mediation of human life with the existence of everything else. When constituted by tissues, the demarcations between inside and outside obtain a distinct spatial quality. The human skin is the largest organ in the body. It separates and holds together the innards, but it also secretes, transmits, absorbs, and stores, and it is above all a multi-layered, partly elastic surface structure interspersed with veins and pores, tendons, muscles, and sensory receptors. It is populated by millions of other living beings. This symbiotic character is found in everything that is made of tissue in the body: in the intestine, in the lungs, in the stomach, even in the cortex and neocortex—we, as humans, are entities of tissues.

If I were to ascribe a commonality to living beings on this earth, it would be the weave, the web, and the mesh. But it is not only that all living things are based on the qualities of weave, web, mesh. Even more significant is the fact that the weave, web, and mesh unfold micro- and macro-architectures, in which living things come together, meeting beyond the open spaces where those elements that are not capable of coexistence run apart and away from each other.

The potential of weave, web, and mesh lies in more than two dimensions. Weave, web, and mesh embody their own spaces.

After an age of destruction of biological diversity, we humans have the task to protect what is left of life on this planet and to grant it its proprietary right to exist. We should therefore no longer perceive architecture as lost, gray energy, as strategically located waste, but as something that offers a place for the metabolic processes of organisms—something that stimulates, protects, and supports them and that is itself alive rather than only imitating the living.

Weave, web, and mesh can be areas in which other organisms nestle, become native, and organize the permeability and the closure between an inside and an outside through their existence. Freed from supporting and loading, the boundary between that embraced inside and that remaining outside at a distance can incorporate other existences, as a woolen sweater holds the rain, thistles, or moths that feed on it. These accumulations and embeddings are not alien to the fabric, but instead realize the potential of its structure. They can trigger interferences. What matters is discovering the unfolding of new potentials in these incidents. Instead of overrating the reproduction circles of a singular species in the enclosure of the architecture, weave, web, mesh allow for the interweaving of reproduction circles of many living beings. For this purpose, fabric, mesh, web do not have to consist of one single material only.

This new space of existence between what we call an inside and what we define as an outside does not confine itself to folded and kinked planes stretched out on pylons. The flexible threads can be combined with rigid wires or rods to form a structure in which compressive and tensile force paths cross in various ways over and around each other. Such an architecture can also push a thread up hill.

If the concept of a merely supporting, loading, and recessed wall is overcome by notions of weave, web, and mesh, which unfold their own spatial qualities between inside and outside, both for the living and for the storage of its resources, then architecture itself is understood as a structure of growth. It provides a basis for existence, stimulating processes of life and therefore of differentiation, and reaching out into the open, the non-determined. "As designers," architect and communist Hannes Meyer wrote in *Bauhaus und Gesellschaft* (Bauhaus and Society) 1929, "we fulfill the destiny of landscape."

Appendix

Image
Credits

p. 81
© Weißensee School of Art and Design
Berlin/Stefanie Eichler, Juni Sun Neyenhuys

pp. 82–83
Top, bottom and left © Weißensee School of
Art and Design Berlin/Stefanie Eichler, Juni
Sun Neyenhuys
p. 83 right (top and bottom) © Max Planck
Institute of Colloids and Interfaces Potsdam/
Golm, Susann Weichold

pp. 84–87
All images © Weißensee School of Art and
Design Berlin/Xingwen Pan

pp. 88–89
© Weißensee School of Art and Design
Berlin/Samira Akhavan

pp. 90–94
© Weißensee School of Art and Design
Berlin/Nelli Singer

p. 97
Fig. 1 © Weißensee School of Art and Design
Berlin/Madeleine Marquardt
Fig. 2 © A New Kind of Blue

pp. 98–101
Fig. 3–5 © STFI Chemnitz

pp. 103–4
All images © Aqualonis

pp. 105–7
All images © Adaptex/Weißensee School of
Art and Design Berlin

pp. 108–14
All images © Kristina Pfeifer, except for
Fig. 7 © Peter Trachsel

pp. 116–17
All images © Hiroyuki Oki

pp. 118–20
p. 118 right (both images) © OLA – Office for
Living Architecture
p. 118 left (both images) © Ferdinand Ludwig
p. 119 top © OLA – Office for Living
Architecture, bottom © Ferdinando Iannone
p. 120 top © OLA – Office for Living
Architecture, bottom © Ferdinando Iannone

p. 121
Top left photo Michelle Mantel © Matters of
Activity; top right © Natalija Miodragović;
bottom © Michaela Eder, Max Planck
Institute of Colloids and Interfaces Potsdam/
Golm; bottom right © Matters of Activity,
Natalija Miodragović, Daniel Suárez

p. 122
Top left © Weißensee School of Art and
Design Berlin/Nelli Singer/photo Daniel
Suárez
Left (middle) © Matters of Activity
Video-still: Lara Ladik; left (bottom)
© Matters of Activity, Daniel Suárez

p. 123
On the right © Nelli Singer, Natalija
Miodragović
Bottom left © Weißensee School of Art and
Design Berlin/Nelli Singer, Daniel Suárez

pp. 124–27
All images © CITA
except for p. 125 bottom © CITA/Yuliya Sinke

pp. 128–31
All images © ICD_ITKE_IntCDC_University
of Stuttgart

pp. 132–35
All images © Weißensee School of Art and
Design Berlin/Idalene Rapp, Natascha Unger

pp. 136–39
All images © Weißensee School of Art and
Design Berlin/Anne-Kathrin Kühner

pp. 140–48
Fig. 1–3 © Christiane Sauer
Fig. 4–8 photo Michelle Mantel © Matters
of Activity

pp. 149–51
All images © Weißensee School of Art and
Design Berlin/Saskia Buch, Martha Panzer,
Jasmin Sermonet, Clara Santos Thomas;
except for p. 151 top © Weißensee School
of Art and Design Berlin/Jasmin Sermonet,
photo Mareike Stoll

pp. 154–57
All images © Samira Boon

pp. 158–59
All images © Weißensee School of Art and
Design/Bára Finnsdóttir

pp. 160–63
All images © Frankfurt UAS, except for p. 160
and p. 161 bottom © Christoph Lison

pp. 164–65
All images © Bas Princen

p. 166
© Eindhoven University of Technology

p. 167
Top © Eindhoven University of Technology,
photo Baart van Overbeeke
Middle © Eindhoven University of
Technology, photo Baart van Overbeeke
Bottom © Eindhoven University of
Technology, photo Baart van Overbeeke

pp. 168–69
© Eindhoven University of Technology

pp. 170–71
All images © Mark West

pp. 172–73
© 2020 Ernesto Neto, photo Albert Sanchez;
p. 173 bottom © 2020 Ernesto Neto,
photo Jens Ziehe/Thyssen-Bornemisza Art
Contemporary Collection, TBA21

pp. 174–80
All images courtesy of Felecia Davis,
except for Fig. 4–6 © Carnegie Museum of
Art, Charles "Teenie" Harris Archive; and for
Fig. 8 courtesy of Farzaneh Oghazian

pp. 181–87
Fig. 1 The Metropolitan Museum of Art
Gift of the Estate of Benjamin E. Marks, 1967
67.180.4, Creative Commons (CC0)
Fig. 2 © Premiere Vision
Fig. 3–4 photo Elaine Igoe
Fig. 5 © Zeitguised
Fig. 6 © Elena Gučas
Fig. 7 © Coram

pp. 190–95
All images © Petra Gruber,
except for Fig. 3 © GrAB

pp. 196–97
All images © EDEK, Universität Kassel

pp. 198–99
All images © Diana Scherer

p. 200
Courtesy of the artist,
photo Studio Tomás Saraceno
© Tomás Saraceno

pp. 201–3
Courtesy of the artist,
photo Andrea Rossetti
© Tomás Saraceno

p. 203
Courtesy of the artist,
photo Andrea Rossetti
© Tomás Saraceno

pp. 204–11
All images © Svenja Keune

pp. 212–16
Fig. 1–4 see captions
Fig. 5 © saai | Archiv für Architektur und
Ingenieurbau am Karlsruher Institut für
Technologie (KIT), Werkarchiv: Frei Otto

Text Credits

Preface
CHRISTIANE SAUER, MAREIKE STOLL

PRACTICES OF MAKING

Introduction
CHRISTIANE SAUER

The Event of a Fiber
REGINE HENGGE, KARIN KRAUTHAUSEN

Bacterial Loom
BASTIAN BEYER

Aerial Construction
AMMAR MIRJAN, FABIO GRAMAZIO,
MATTHIAS KOHLER

Augmented Spinning
ELAINE BONAVIA, KAROLA DIERICHS

Slip-Form Rock Jamming
MAXIE SCHNEIDER

Woven Paper Bridge
LORENZO GUIDUCCI, MAXIE SCHNEIDER,
JOSEPHINE SHONE

Living Root Bridges
WILFRID MIDDLETON,
FERDINAND LUDWIG

Dorze Architecture
PETRA GRUBER

Topologies of Weaving
EBBA FRANSÉN WALDHÖR

DESIGNING PERFORMANCE

Introduction
CHRISTIANE SAUER

Plant Fiber Twists
MICHAELA EDER, MARTIN NIEDERMEIER,
CHARLETT WENIG

Biomaterials and Design
PETER FRATZL, CHRISTIANE SAUER

Interweaving Disciplines
LORENZO GUIDUCCI

Hydroweave
CHRISTIANE SAUER

Skeletal Surface
XINGWEN PAN, CHRISTIANE SAUER

Weaving Wrinkles
EBBA FRANSÉN WALDHÖR

Living Beings
NELLI SINGER, CHRISTIANE SAUER

Technical Textiles
CHRISTIANE SAUER,
HEIKE ILLING-GÜNTHER

CloudFisher®
MAXIE SCHNEIDER

Adaptex
EBBA FRANSÉN WALDHÖR,
MAXIE SCHNEIDER

Black Hair Tents
KRISTINA PFEIFER

FIBER STRUCTURES

Introduction
CHRISTIANE SAUER

Vedana Restaurant
MAXIE SCHNEIDER

Plane Tree Cube
FERDINAND LUDWIG, DANIEL SCHÖNLE,
JAKOB RAUSCHER

Structural Textile
NATALIJA MIODRAGOVIĆ

Hybrid Tower
MARTIN TAMKE, METTE RAMSGAARD
THOMSEN, YULIYA SINKE
BARANOVSKAYA

Maison Fibre
EBBA FRANSÉN WALDHÖR

Stone Web
IDALENE RAPP, NATASCHA UNGER,
CHRISTIANE SAUER

Concrete Textile
ANNE-KATHRIN KÜHNER,
CHRISTIANE SAUER

Architectural Yarns
IVA REŠETAR, CHRISTIANE SAUER

Scaling Fiber
SASKIA BUCH, MARTHA PANZER,
JASMIN SERMONET, CLARA SANTOS
THOMAS, IVA REŠETAR,
CHRISTIANE SAUER

FABRICATING SPACE

Introduction
CHRISTIANE SAUER

ArchiFolds
EBBA FRANSÉN WALDHÖR

Merging Loops
BÁRA FINNSDÓTTIR, CHRISTIANE SAUER

SpacerFabric_Pavilion
CLAUDIA LÜLING

Centres for Traditional Music
EBBA FRANSÉN WALDHÖR

Sagrada Familia in Ice
ARNO PRONK

Fabric Formwork
MAXIE SCHNEIDER

SunForceOceanLife
EBBA FRANSÉN WALDHÖR

Fabricating Networks Quilt
FELECIA DAVIS

Textile Fenestra and Spiracles
ELAINE IGOE

ENTANGLED ENVIRONMENTS

Introduction
CHRISTIANE SAUER

Biological Entanglements
PETRA GRUBER

Home
MAXIE SCHNEIDER

Interwoven
EBBA FRANSÉN WALDHÖR

Webs of At-tent(s)ion
EBBA FRANSÉN WALDHÖR

Multispecies Cohabitation
SVENJA KEUNE, DELIA DUMITRESCU,
METTE RAMSGAARD THOMSEN

Living with Moths
JÖRG PETRUSCHAT

Contributors

EDITORIAL TEAM

Prof. CHRISTIANE SAUER is an architect and Professor for Material Design in Architectural Context at Weißensee School of Art and Design Berlin. Her focus is on novel and sustainable material developments for the architectural context based on textile structures, active materials and functional surfaces. In her work she bridges experimental research and design practice. She co-heads the research facility DXM — Design Experiment Material and is Principal Investigator at the Cluster of Excellence *Matters of Activity. Image Space Material*. As a lecturer and architect, she is working internationally in an increasingly interdisciplinary context.

Dr. MAREIKE STOLL obtained her PhD in German Studies from Princeton University, defending her dissertation on photobooks of the Weimar Period in 2015. At Weißensee School of Art and Design Berlin she is currently a Research Associate at the Cluster of Excellence *Matters of Activity. Image Space Material*, where she facilitates the exchange of interdisciplinary research. In her independent research, she investigates images in books — from ABC books to the atlas, from picture books to photobooks — and how all the elements in the book constitute a constellation of epistemological inquiry and exchange.

EBBA FRANSÉN WALDHÖR is an artist, designer, and lecturer who works in spatial installation and design research, with a primary focus on textiles. As a Design Researcher at the Weißensee School of Art and Design Berlin, she explores adaptive materials in textiles for architectural contexts. In her design practice, she develops experimental spatial concepts and scenographies for artists and institutions. The physicality of textiles, as soft barriers, provides a material ground from which she examines questions of spatial boundaries and their permeability. She teaches Scenography and Exhibition Design at the University of Arts and Design in Karlsruhe.

MAXIE SCHNEIDER is an architectural design researcher. Her work combines physical and digital prototyping to develop new building techniques and material systems. She has collaborated on various design-build projects and advanced material experiments into structural implementation. As a Research Associate and PhD candidate at the Cluster of Excellence *Matters of Activity. Image Space Material* and TU Berlin, she investigates adaptive textile hybrid structures and aspects of the functionalization of softness in architecture. She teaches textiles in spatial context at Weißensee School of Art and Design Berlin.

PRACTICES OF MAKING

Prof. Dr. REGINE HENGGE studied Biology and obtained her doctorate at Universität Konstanz. After a postdoctoral phase at Princeton University (US), she completed her Habilitation at Universität Konstanz. She has been a Full Professor of Microbiology at Freie Universität Berlin (1998–2013) and Humboldt-Universität zu Berlin (since 2013). She studies signal transduction networks and gene regulation in stationary phase bacteria and emergent properties of multicellularity in bacterial biofilms. She received the Gottfried Wilhelm Leibniz Prize and an ERC Advanced Investigator Grant. She is an elected member of the German National Academy Leopoldina, EMBO and the American Academy of Microbiology. In the Cluster *Matters of Activity. Image Space Material*, she is Principal Investigator in the project *Weaving*.

Dr. KARIN KRAUTHAUSEN is a Research Associate in the Cluster *Matters of Activity. Image Space Material* at Humboldt Universität zu Berlin where she works as a literary scholar and cultural historian in the project *Weaving*. Her present research focus is design procedures in the arts and sciences, the relation of realism and structuralism in German literature, and the poetics of history. Recent publications: Peter Fratzl, Michael Friedman, Karin Krauthausen, Wolfgang Schäffner (ed.), *Active Materials* (De Gruyter, 2021); Karin Krauthausen, Rebekka Ladewig (ed.), *Modell Hütte. Von emergenten Strukturen, schützender Haut und gebauter Umwelt* (Diaphanes, 2021).

Dr. BASTIAN BEYER's research deals with fiber-based materials in the architectural context, with a special focus on new manufacturing methods for bio-based materials. Dr. SKANDER HATHROUBI is a microbiologist and microbial biofilm researcher. He is particularly interested in new tools to study morphogenetic movements in macro-colony biofilm, an accessible model for active matter. Together with Prof. Dr. REGINE HENGGE, they are working on cellulose-based biofilms and their application for design and architecture. They are both Research Associates in the Cluster *Matters of Activity. Image Space Material* in the research project *Weaving*.

GRAMAZIO KOHLER RESEARCH, led by Prof. FABIO GRAMAZIO and Prof. MATTHIAS KOHLER is based at ETH Zurich and examines the changes in architectural production requirements that result from introducing digital manufacturing techniques. Opening the world's first architectural robotic laboratory, their research has been formative in the field of digital architecture, setting precedence and de facto creating a new research field merging advanced architectural design and additive fabrication processes. Dr. AMMAR MIRJAN leads a research group that examines multi-robotic assembly processes and their implication on architectural design.

Prof. Dr. KAROLA DIERICHS holds the Professorship Material and Code at Weißensee School of Art and Design Berlin as part of the Cluster of Excellence *Matters of Activity. Image Space Material* at Humboldt-Universität zu Berlin. Previously she was a Research Associate at the Institute for Computational Design and Construction (ICD) within the Cluster of Excellence Integrative Computational Design and Construction for Architecture (IntCDC). ELAINE BONAVIA is an architectural design researcher with a background in computational design and engineering. Her research interest is in elastic architecture, soft spaces, and movement-body-space relationships across physical and virtual dimensions. Since 2020 she has been teaching in the Department of Textile and Surface Design at Weißensee School of Art and Design Berlin.

The Self-Assembly Lab, founded by Prof. SKYLAR TIBBITS and co-directed with JARED LAUCKS, is located at MIT's cross-disciplinary International Design Center. The center advances knowledge in the field of self-assembly and programmable material technologies. Prof. SKYLAR TIBBITS is Associate Professor of Design Research in the Department of Architecture at MIT. He has designed large-scale installations and exhibited in galleries including Centre Pompidou, Paris, and Victoria and Albert Museum, London. JARED LAUCKS is a Research Scientist and holds a Master of Science from MIT Media Lab where he focused on robotic and biological fabrication methods.

JOSEPHINE SHONE completed her BA in Textile and Surface Design at Weißensee School of Art and Design Berlin. Her interest is in combining traditional craft techniques and modern technologies to develop tools and techniques that reuse, repurpose and recycle. At the Cluster of Excellence *Matters of Activity. Image Space Material*, she collaborates with Dr. LORENZO GUIDUCCI, IVA REŠETAR, MAXIE SCHNEIDER and Prof. CHRISTIANE SAUER to explore friction-based materials and Architectural Yarns.

As part of the research field Baubotanik (living plant construction), WILFRID MIDDLETON and Prof. Dr. FERDINAND LUDWIG are researching Indian root bridges at Technical University Munich. Prof. Dr. FERDINAND LUDWIG holds the Chair for Green Technologies in Landscape Architecture where WILFRID MIDDLETON is researching Living Root Bridges as part of his dissertation. The term Baubotanik was first developed at the Institute for Modern Architecture and Design (IGMA) at the University of Stuttgart. It blends the fields of research and development and is a method of construction that utilizes living plants as load-bearing systems in architectural structures.

Prof. Dr. MYFANWY EVANS is a mathematician and physicist working on the geometric and topological modeling of soft and biological systems. She is Professor for Applied Geometry and Topology in the Institute for Mathematics at the University of Potsdam. She completed her PhD in Mathematics at the Australian National University, and between 2015 and 2020, she led an Emmy Noether Research Group at the Technical University, Berlin, on Geometry and Topology of Entangled Soft Matter. She is Principal Investigator in the Cluster *Matters of Activity. Image Space Material* at Humboldt Universität zu Berlin in the project *Weaving*.

ALISON GRACE MARTIN is an artist, weaver and independent researcher creating complex geometric shapes, primarily in paper and bamboo. She studied at Exeter College of Art and St. Martin's College of Art and Design, graduating in Graphic Design and Visual Communication in 1979. Her work involves the analysis of geometry and topology in traditional weaving patterns that lead to light and flexible designs, at all scales. She lives and works in Fivizzano, Italy.

DESIGNING PERFORMANCE

Adaptive Fibrous Materials is a Research Group located at the Max Planck Institute of Colloids and Interfaces, Department of Biomaterials, Potsdam-Golm and the Cluster of Excellence *Matters of Activity. Image Space Material*. Led by wood scientist Dr. MICHAELA EDER they research interactions between biological materials and their environments and combine materials-science tools at various length scales with design experiments. They work towards a deep understanding of material activity, adaptation, and optimization strategies, and explore material properties under consideration of developmental stages (growth) and environmental conditions in the field or in the lab. Biologist MARTIN NIEDERMEIER and product designer CHARLETT WENIG currently join the team as Pre-Doctoral Researchers.

Prof. Dr. Dr. h.c. PETER FRATZL is director at the Max Planck Institute of Colloids and Interfaces in Potsdam, heading the Department of Biomaterials. He holds an engineering degree from Ecole Polytechnique Paris, a doctorate in Physics from Universität Wien and is honorary Professor at University of Potsdam and Humboldt-Universität zu Berlin, where he co-directs the Cluster of Excellence *Matters of Activity. Image Space Material*. His research is in interdisciplinary materials science, focusing on biological and bioinspired materials. He is recipient of the Leibniz Award and member of several Academies of Science in Berlin, Mainz, and Austria, as well as of the German Academy of Engineering and the US National Academy of Engineering.

Dr. LORENZO GUIDUCCI is a biomedical engineer with a PhD in Physics, Science of Biomaterials. In his research he combines computational simulations, theoretical modeling, and tabletop experiments to design and understand structure-function relationships in bioinspired architected materials. His work and interests often overlap with those from applied designers with whom he collaborates. At present, he is Research Associate in the *Material Form Function* project in the interdisciplinary Cluster of Excellence *Matters of Activity. Image Space Material* at Humboldt-Universität zu Berlin.

JUNI SUN NEYENHUYS has a background in Textile and Surface Design as well as Product Design, and works at the intersection of innovative, sustainable material development and conceptual design. During her studies, she specialized in the development of algae-based materials. In 2021, she co-founded the start-up mujò, which develops algae-based, biodegradable packaging materials. STEFANIE EICHLER focused her studies in Textile and Surface Design on developing sustainable design approaches, combining traditional craft techniques with renewable resources. Since 2022, she has been working as a textile designer in her own Berlin-based studio.

XINGWEN PAN currently studies at the Master Study Program at Textile and Surface Design at Weißensee School of Art and Design Berlin. Her special interest lies in the connection between materials and technology. She works on form-finding processes based on material properties together with visual and tactile aspects.

SAMIRA AKHAVAN is a product and textile designer based in Berlin. Her work deals with the construction, aesthetics, and functions of surfaces. She graduated from the Department of Textile and Surface Design at Weißensee School of Art and Design in 2021 with her Master Thesis *Woven Systems* and is currently an academic employee at University of Applied Sciences, Potsdam.

NELLI SINGER is a textile designer and material researcher with a strong interest in biomaterial activity. The technology used for her Master Thesis *Living Beings* became a central part of the Active Curtain, an interdisciplinary exhibit of the Cluster of Excellence *Matters of Activity. Image Space Material* at the Humboldt Lab, Berlin. In the Cluster she closely collaborated on the development of Active Curtain and Structural Textile with NATALIJA MIODRAGOVIĆ, DANIEL SUAREZ and Prof. CHRISTANE SAUER.

Dr. HEIKE ILLING-GÜNTHER has a degree in Engineering Chemistry and a doctorate in the field of Food and Environmental Toxicology. She entered textile testing and research in 1996 at the Textile Research Institute Thuringia-Vogtland Greiz. Since 2006 she has been working at the Saxon

Textile Research Institute (STFI) Chemnitz. From 2010 to 2021 Illing-Günther was the Research Director of STFI, since 2021 she has been Managing Director. Her key activities focus on research and development of Technical Textiles and Smart Textiles.

PETER TRAUTWEIN is an industrial designer who develops, distributes and installs CloudFisher® fog collectors worldwide with his company Aqualonis GmbH in collaboration with the WaterFoundation. Trautwein holds a degree in industrial design and has been working in the fields of transportation design, consumer electronics, medical technology, exhibitions and water treatment since 1989.

DXM—DESIGN EXPERIMENT MATERIAL is the research group of the Department of Textile and Surface Design at Weißensee School of Art and Design and co-headed by Prof. CHRISTANE SAUER who works together with MAXIE SCHNEIDER and EBBA FRANSÉN WALDHÖR on the research, design and development of Adaptex. PRIEDEMANN FACADE-LAB GmbH, Großbeeren, is an international facade-engineering company, co-headed by PAUL-ROUVEN DENZ, who is an architect with a focus on Smart Textile Skin solutions for material and energy-efficient building envelopes. He works together with architects PUTTAKHUN VONGSINGHA, NATCHAI SUWANNAPRUK and Dr. JENS BÖKE on the engineering concept and development of Adaptex.

Dr. KRISTINA PFEIFER studied architecture at the Technical University Vienna, where her doctoral thesis in 2015 led her to continuous work in her specialty: black hair tents. Since 2004 she has undertaken several field-research tasks in Turkey and Morocco, published a documentary movie, conducted laboratory tests, held academic workshops and lectures on the subject, and supported contacts between scholars and Yörük nomads. Currently, she is involved in a research project about yak hair tents on the Tibetan Plateau as a co-author, and in articles for an encyclopedia on vernacular architecture.

FIBER STRUCTURES

VTN ARCHITECTS was founded in 2006 by VO TRONG NGHIA in Ho Chi Minh City, Vietnam. The office incorporates traditional Vietnamese building techniques like perforated blocks, cooling water systems, shaded terraces, thatched roofs, and local materials for its innovative approach based on vernacular building tradition.

Prof. Dr. FERDINAND LUDWIG is a pioneering architect in the field of Baubotanik and Professor for Green Technologies in Landscape Architecture at Technical University Munich. Together with his partners DANIEL SCHÖNLE and JAKOB RAUSCHER he heads OLA—Office for Living Architecture in Stuttgart, where they develop novel forms of building typologies by making use of the structures and growth processes of living trees that redefine the relation between nature and technology.

NATALIJA MIODRAGOVIČ is an architect and Research Associate in the projects Object Space Agency and Weaving at the Cluster of Excellence *Matters of Activity. Image Space Material*. The focus of her interdisciplinary and experimental work is art and space as vehicles for social change. DANIEL SUÁREZ is an architect and Associated Member of the Cluster who operates at the intersection of textile technology and architecture. Together with Dr. BASTIAN BEYER, he investigates the biocalcification of textile scaffolds. In

the ongoing research project Structural Textile, NATALIJA MIODRAGOVIĆ, DANIEL SUÁREZ, and NELLI SINGER, together with Prof. CHRISTIANE SAUER explore the potential of the inner structural activity of plant-fiber for scaffolds and bio-composites on an architectural scale through bridging craft and computation.

Prof. Dr. METTE RAMSGAARD THOMSEN is head of CITA—Centre for Information Technology and Architecture at the Royal Danish Academy of Fine Arts, School of Architecture, Design, Conservation in Copenhagen. Prof. MARTIN TAMKE is Associate Professor in the field of digital production and architectural computation. YULIYA SINKE BARANOVSKAYA is an architectural researcher specializing in CNC textile technologies. At CITA they investigate and teach advanced computational modeling, digital fabrication and material specification using a practice-based research method, focused on the conceptualization, design, and realization of working prototypes and full-scale demonstrators.

The Institute for Computational Design and Construction (ICD) at the University of Stuttgart is directed by Prof. ACHIM MENGES. His focus is on the development of integrative design at the intersection of computational design methods, robotic manufacturing and construction, as well as advanced material and building systems. Prof. Dr. JAN KNIPPERS is director of the Institute of Building Structures and Structural Design (ITKE) with a focus on integrating computational engineering and advanced analysis methods together with fabrication and development of full-scale architectural prototypes. Achim Menges is director and Jan Knippers is a member of the board of directors of the Cluster of Excellence Integrative Computational Design and Construction for Architecture (IntCDC), University of Stuttgart.

IDALENE RAPP and NATASCHA UNGER graduated from the Department of Textile and Surface Design at Weißensee School of Art and Design. They are a Berlin-based experimental design duo working together as RappUnger. Their common interest is driven by exploration and experimentation through spatial design and materiality with a portfolio ranging from everyday objects to unconventional installations.

ANNE-KATHRIN KÜHNER developed Concrete Textile as her Master Thesis at Weißensee School of Art and Design Berlin. She is a textile and material designer with a strong focus on textile techniques and materials. The emphasis of her work is on the development of large-scale textiles that can be adapted in their construction, materiality, and dimensions.

IVA REŠETAR studied architecture at the University of Belgrade and at the Städelschule in Frankfurt am Main. She has been a practicing architect in the field of digital and experimental design, and has held several teaching and research positions, among others at the Akademie Schloss Solitude, Weißensee School of Art and Design and at the Berlin University of the Arts. In her research, she examines the relationships between materials and thermal environments, involving experiments with fibrous architectures. Rešetar has been Research Associate at the Cluster of Excellence *Matters of Activity. Image Space Material* at Humboldt-Universität zu Berlin since 2019.

SASKIA BUCH, MARTHA PANZER, JASMIN SERMONET, and CLARA SANTOS THOMAS currently study Textile and Surface Design at Weißensee School of Art and Design Berlin. In the Design Research Studio Scaling Fiber they worked together as a team on the design and making of Experimental Yarns, with Saskia Buch focusing on wool, Martha Panzer on textile waste and Jasmin Sermonet on earth as yarn materials.

FABRICATING SPACE

Studio Samira Boon is led by architect SAMIRA BOON and currently based in Amsterdam and Tokyo. The studio specializes in combining the sensory qualities of textile materials with research into state-of-the-art computerized production techniques. As an expert in the material properties of textiles, Studio Samira Boon closely collaborates with architects to design site-specific adaptive and dynamic solutions.

BÁRA FINNSDÓTTIR is an Icelandic designer, living in Berlin. She graduated in 2018 in Textile and Surface Design from the Weißensee School of Art and Design. Her projects are characterized by experimental material studies in which materials are explored through novel processing or construction. Surfaces designed for an architectural context as well as sustainable, nature- and science-based concepts are dominant in her work.

Prof. CLAUDIA LÜLING is Professor at the Frankfurt University of Applied Sciences where she conducts research on textile-based lightweight construction. One focus is the development of sustainable components made of technical 3D spacer textiles in combination with 3D printing and filling techniques. For her work, she was recently awarded an innovation professorship by Frankfurt UAS, among others.

OFFICE Kersten Geers David Van Severen, based in Brussels, was founded in 2002 by KERSTEN GEERS and DAVID VAN SEVEREN. Their renowned practice is known for its minimalist yet idiosyncratic architecture, both in realizations and in theoretical projects. The firm reduces architecture to its very essence and most original form: a limited set of basic geometric rules is used to create a framework from which life unfolds in all its complexity.

Prof. ARNO PRONK is Assistant Professor at Eindhoven University of Technology, where he leads the research group Structural Ice, which has realized a series of remarkable structures made from composite ice material. His book on the topic was published in 2021: *Flexible Forming for Fluid Architecture* (Cham: Springer International Publishing, 2021). His expertise is in the development and application of environmentally related innovations in experimental structures and buildings.

Prof. MARK WEST founded the Centre for Architectural Structures and Technology (C.A.S.T.) at the University of Manitoba, Canada, in 1998, as the first university laboratory dedicated to textile formwork technology. He has taught architecture at a number of universities throughout North America, while working as an artist, inventor, and independent researcher. Since 2017, West has been focusing on visual explorations of the topic in his studio "Surviving Logic" in Montreal.

ERNESTO NETO is an artist who lives and works in Rio de Janeiro. He studied sculpture at the Escola de Artes Visuais do Parque Lage in Rio de Janeiro. His textile installations fill exhibition spaces with biomorphic environments that invite social and physical interaction. Neto's work is represented in international museum collections, including the Museum of Modern Art in New York, Tate Gallery in London, Centre Pompidou in Paris or Hara Museum in Tokyo among many others.

Prof. Dr. FELECIA DAVIS is an Associate Professor at the School of Architecture at Pennsylvania State University and is the director of SOFTLAB@PSU. She completed her PhD in Design Computation at MIT. Davis' work in architecture connects art, science, engineering, and was featured by PBS in the Women in Science Profiles series. Her work was part of MoMA's exhibition Reconstruction: Blackness and Architecture in America. She is a founding member of the Black Reconstruction Collective supporting design work about the Black diaspora. Davis' work has been awarded by the New York Architectural Leagues' 2022 Emerging Voices in Architecture program.

Dr. ELAINE IGOE is Senior Lecturer in Textile Design at Chelsea College of Arts, University of the Arts London and has taught and coordinated undergraduate, postgraduate, and doctoral students at the University of Portsmouth and the Royal College of Art. Igoe has made key contributions to the critical theory of textile design, applying an approach that interrogates the discipline of textiles from within while simultaneously addressing theoretical discourses and wider contexts.

ENTANGLED ENVIRONMENTS

Dr. PETRA GRUBER is an architect with a passion for biology and biomimetic design. She holds a PhD in Biomimetics in Architecture from Vienna University of Technology in Austria and has worked internationally on three continents in inter- and transdisciplinary design, research, and education, at the intersection of biology, architecture, and art. Her work on spatial and functional aspects of biological structures for biomimetic innovation in architecture and the built environment has been published widely. Currently she is an expert at FFG, the Austrian Research Promotion Agency.

In the research project HOME, the Department of Experimental and Digital Design and Construction (EDEK) at the University of Kassel, led by Prof. PHILIPP EVERSMANN, was responsible for organization, architectural design, and additive manufacturing process. Prof. Dr. JAN WURM led the Arup team for circular design, structural and acoustic engineering, and parametric modeling. The team of Prof. DIRK HEBEL, at the Department of Architecture of KIT, supervised the bio-cultivation process, the mycelium characterization, and the structural testing of components.

DIANA SCHERER's art practice encompasses botany, photography, and sculpture based on collaborations with biologists and engineers. She studied Fine Art at the Gerrit Rietveld Academy in Amsterdam, where she continues to live and work. Over the past years she has exhibited in several international solo and group shows.

TOMÁS SARACENO is an Argentina-born, Berlin-based artist whose projects dialogue with different forms of life and life-forming, rethinking dominant threads of knowledge in the Capitalocene. Studio Tomás Saraceno was founded in 2005 by the artist. Based in Berlin, the studio is composed of members with multidisciplinary backgrounds, including designers, architects, anthropologists, biologists, engineers, art historians, curators, and musicians. Saraceno's work has been the subject of solo exhibitions and permanent installations at museums and institutions internationally, including Palais de Tokyo, Paris, the Metropolitan Museum of Art, New York, and Hamburger Bahnhof – Museum für Gegenwart, Berlin.

Dr. SVENJA KEUNE is a postdoctoral researcher at the Swedish School of Textiles and at the Centre for Information Technology and Architecture (CITA) at the Royal Danish Academy in Copenhagen collaborating with Prof. Dr. METTE RAMSGAARD THOMSEN. During her PhD project On Textile Farming within the MSCA ArcInTexETN she turned towards seeds as a dynamic material for textile design and moved into a Tiny House on wheels to live together with her research experiments. She is currently working on Designing and Living with Organisms (DLO), a three-year project funded by an international postdoctoral grant from the Swedish Research Council.

Prof. Dr. DELIA DUMITRESCU is a Professor in Textile Design at the Swedish School of Textiles and Head of the Smart Textiles Design Lab, Science Park Borås. Her research focuses on the development of Smart Textiles, design methods, and aesthetics, using industrial textile manufacturing and digital technology for diverse applications, from the body to interiors. She was the Director of Studies for ArcintexETN international research school granted by the EU Horizon 2020, MSC Actions, where fifteen PhD students conducted cross-disciplinary research in the area of textiles, interaction, and architectural design.

Prof. Dr. JÖRG PETRUSCHAT has been Professor for Theory and History of Design at Weißensee School of Art and Design Berlin since 2014. His research topics are the ability to design and its cognitive dimensions, functions of form in evolution, Gestaltung, and performative research. He studied philosophical aesthetics, cultural sciences, history of art, design and urbanism. Petruschat was research assistant at the Institute for Aesthetics of the Humboldt-Universität zu Berlin with Günter Mayer and Friedrich Kittler from 1986 to 1998. Between 1991 to 2008 he was Editor-in-Chief of the magazine *Form+Zweck. Zeitschrift für Gestaltung*, and since 2019 he has been head of publishing. He is Associate Investigator in the Cluster of Excellence *Matters of Activity. Image Space Material*.

Imprint

Editorial team
CHRISTIANE SAUER
MAREIKE STOLL
EBBA FRANSÉN WALDHÖR
MAXIE SCHNEIDER

The editors would like to acknowledge the
support of the Cluster of Excellence *Matters
of Activity. Image Space Material* funded
by the Deutsche Forschungsgemeinschaft
(DFG, German Research Foundation)
under Germany's Excellence Strategy —
EXC 2025 — 390648296 (funding period
1 January 2019 — 31 December 2025).

A special thank you goes to everyone who
made this book possible by providing help
and advice in administration, financing, and
content as well as to family and friends for
their enduring patience and support during
the intense phase of production.

Every effort has been made to supply complete and correct credits. In case of any er-
rors or omissions, please contact the editors
so that corrections can be incorporated into
subsequent editions of this publication.

Copy editing
MELISSA LARNER

Design and setting
FLOYD E. SCHULZE

Cover
front including inside cover: *Stone Web* ©
Weißensee School of Art and Design Berlin/
IDALENE RAPP, NATASCHA UNGER;
back including inside cover: *Dorze
Architecture* © PETRA GRUBER

Production
SUSANNE RÖSLER

Lithography
BILD1DRUCK, Berlin

Printed in the EUROPEAN UNION

Bibliographic information published by
the Deutsche Nationalbibliothek
The Deutsche Nationalbibliothek
lists this publication in the Deutsche
Nationalbibliografie; detailed bibliographic
data are available on the Internet at
http://dnb.d-nb.de

jovis Verlag GmbH
Lützowstraße 33
10785 Berlin

www.jovis.de

jovis books are available worldwide in select
bookstores. Please contact your nearest
bookseller or visit www.jovis.de for information concerning your local distribution.

ISBN 978-3-86859-739-4 (Softcover)
ISBN 978-3-86859-831-5 (PDF)